D1356078

Bernhard Steffen • Oliver Rüthing • Michael Huth

Mathematical Foundations of Advanced Informatics

Volume 1: Inductive Approaches

 Springer

Bernhard Steffen
Fakultät für Informatik
Technical University of Dortmund
Dortmund, Nordrhein-Westfalen
Germany

Oliver Rüthing
Fakultät für Informatik
Technical University of Dortmund
Dortmund, Nordrhein-Westfalen
Germany

Michael Huth
Department of Computing
Imperial College London
London, United Kingdom

ISBN 978-3-319-68396-6   ISBN 978-3-319-68397-3   (eBook)
https://doi.org/10.1007/978-3-319-68397-3

Library of Congress Control Number: 2018933380

Printed on acid-free paper

This Springer imprint is published by the registered company Springer International Publishing AG part of Springer Nature.
The registered company address is: Gewerbestrasse 11, 6330 Cham, Switzerland

We dedicate this book to all those future Informaticians who will seek – with determination – points on which they can stand quiescently.

Foreword

Mathematical logic was developed in an attempt to confront the crisis in the foundations of mathematics that emerged around the turn of the 20th century. Between 1900 and 1930, this development was spearheaded by Hilbert's Program, whose main aim was to formalize all of mathematics and establish that mathematics is complete and decidable. Hilbert firmly believed that these ambitious goals could be achieved. Nonetheless, Hilbert's Program was dealt devastating blows during the 1930s. Indeed, the standard first-order axioms of arithmetic were shown to be incomplete by Gödel in his celebrated 1931 paper. Furthermore, Turing, Church, and Tarski demonstrated the undecidability of first-order logic. Specifically, the set of all valid first-order sentences was shown to be undecidable, whereas the set of all first-order sentences that are true in arithmetic was shown to be highly undecidable.

Today, mathematical logic is a mature and highly sophisticated research area with deep results and a number of applications in certain areas of mathematics. All in all, however, it is fair to say that the interaction between logic and mathematics has been somewhat limited. In particular, mathematical logic is not perceived as one of the mainstream areas of mathematics, and the typical mathematician usually knows little about logic. In contrast, logic has permeated through computer science during the past 50 years much more than it has through mathematics during the past 100 years. Indeed, during the past 50 years there has been extensive, continuous, and growing interaction between logic and computer science. In many respects, logic provides computer science with both a unifying foundational framework and a tool for modeling computational systems. In fact, concepts and methods of logic occupy such a central place in computer science that logic has been called "the calculus of computer science".

This trilogy lays out the logical foundations of informatics in a comprehensive and systematic way. It first develops the mathematical foundations of propositional and first-order logic, as well as its primary proof technique, the inductive method. It then develops a logical approach to algebraic thinking, which enables the reader to see the logical commonality in diverse algebraic structure. It concludes with the detailed development of a specific algebraic structure, linear algebra, and shows how it can be used for both modeling and analysis.

The reader who pursues the foundations laid out in this trilogy is guaranteed to develop both a deep appreciation of these foundations, as well as the capacity to use these foundations in diverse applications in computing.

Houston, May 2017   *Moshe Y. Vardi*

Prologue

Informatics is a cross-cutting discipline that sits between the disciplines of Mathematics, the Engineering Sciences, and even the Business Sciences. As such it relates to disciplines that reflect very different cultures of research and development. How do such differences manifest themselves? We would like to provide anecdotal evidence based on personal conversations that the first author has had. The summaries of the example conversations below make clear how well running jokes that compare cultures of different disciplines often capture such differences:

A Mathematician was asked, at the beginning of the 1990s, whether he thought that area-wide mobile telephony would be possible. The given answer was: *"This is impossible, as it would require the installation of millions of aerial masts."*

An Engineer, when asked the same question, answered: *"We will see such systems in the future, since it will only require the installation of a few million aerial masts."*

In the second scenario, an Informatician was at a project meeting and asked by a client whether half a man-year would be adequate for the realization and completion of a specific technical task. Upon which, the Informatician exclaimed: *"But the underlying problem of this task is undecidable!"* The client then quickly reflected on that statement and expressed his willingness to extend the resource allocation from half a man-year to an entire man-year.

Today, nobody needs to explain the success of mobile telephony. And it is astonishing how many problems that are undecidable in theory have the status of "successful, working implementations" in practice. Why do we have such apparent contradictions, and how should a budding Informatician deal with them? Put in another way: what can we learn from this, and what influence do such observations have on the relevance of Mathematics in practice? Or, to put it more provocatively: When faced with such observations, is Mathematics still significant for Informatics?

Mathematicians are trained to be pessimists. They always assume the worst and only accept something as being proven if it cannot be false under any circumstances. Such a line of thinking is also reflected in those areas of Informatics that are strongly shaped by mathematical thinking. For example, in complexity theory we typically study the *worst-case* behavior of an algorithm. The prominence of worst-case analysis is not only rooted in a pessimistic approach, though. It is also a reflection of the fact that more practically relevant notions such as *average*-case complexity are much harder to develop and apply to algorithms.

On the other hand, an Engineer may at times realize a solution without penetrating the full scope of the underlying problem. This may lead to the implementation of an algorithm that superficially seems to be simple to implement but whose implementation may fail to terminate. Nevertheless, a pragmatic engineering approach can create truly astonishing success, and can produce a sense of achievement that may not be attainable through complexity-theoretic considerations. An example thereof is the well known Simplex Method for linear programming, whose worst-case complexity is exponential but which in practice outperforms methods whose worst-case complexity is proved to be polynomial. Another example is the decision procedure for second-order monadic logic, which has non-elementary worst-case complexity.[1] For about 30 years, this was seen to mean that there is no point in implementing such an algorithm for practical use. But we now have such implementations of practical value.

This context makes it not easy for a practically minded Informatician who embraces conceptual thinking to position herself. But it opens up a new perspective and approach. It also awakens the desire to understand the reasons behind the discrepancies outlined above, and whether we might be able to explain the successes of engineering-based Informatics. To put it in mathematical form: what are the sufficient and the necessary conditions for characterizing success scenarios? In fact, we are only concered with sufficient conditions when we see that something is working really well, and we would like to be able to generalize and re-apply the reasons for such success. It is important that such re-applications secure future success, and this also informs how general success criteria should be: a user is not interested in details of criteria that are irrelevant to obtaining success for the problem at hand.

This constructive approach to thinking and problem solving begins with the consideration of something that functions, is certain, or is efficient. Then it attempts to generalize, step by step, in a manner that is driven by the needs of the new problem to solve. We may think of this approach as being complementary to the classical approach in Mathematics, which aims to achieve maximal generality for both definitions and proven facts in order to maximize their reach. The constructive approach also uses such a methodological and mathematical approach, yet with different priorities than those found in Pure Mathematics.

The aim of this trilogy is to capture the core of Advanced Informatics by making its required foundations accessible and to enable students to become effective problem solvers [44] who are also "already eating" in the sense of the joke seen in the footnote, which is not meant to pay disrespect to particular disciplines.[2]

---

[1] This means that the worst-case complexity for input parameter n cannot be bounded above by something that grows asymptocically like $(\cdots((2^2)^2)^2\cdots)^n)$ for any finite number of nestings in $\cdots$.

[2] A Mathematician, a Physicist, and an Engineer are confined in a closed room and each one of them has a large tin can filled with ravioli at their disposal but no tool for opening it. After a few

---

hours of inspection, the Mathematician sits in front of his tin can with closed eyes and says "Let the tin can be open," whereas the Physicist is still busy with calculating the optimal angle for throwing the tin can against the wall. What is the Engineer doing? He is eating already!

Preface

Advanced Informatics – What Is That Anyway?

The title of this trilogy is inspired by the term *Advanced Mathematics*, which refers to the competent use, at university level, of Mathematics in the natural and engineering sciences. This competency goes beyond the ability to follow numerical recipes and perform calculations. But it neither requires the mastery of complex proofs nor a thorough understanding of recent mathematical results. By analogy, we may refer to *Advanced Informatics* as a conceptual pragmatics that concerns itself with the notion of information, and which goes beyond the core competencies of writing programs and scripts, and clicking through applications. Advanced Informatics – as we understand it here – does not require the mastery of theoretical research in Informatics and therefore has a more applied outlook. The supporting theoretical foundations, however, belong to its taught curriculum just as is the case for Advanced Mathematics. The content of this trilogy does provide such Theoretical Informatics foundations, which compare in terms of complexity and importance to theoretical foundations of Mathematics such as the mastery of infinities through convergence criteria or the characterization of essential structure through axiomatizations.

A key characteristic of Advanced Informatics is that it offers general patterns of thought and process that provide crucial aids and guidelines in dealing with the sheer unlimited possibilities and degrees of freedom in computer-supported solutions to real-world problems. In this sense, Advanced Informatics goes beyond the scope of Mathematics in that it not only develops methodologies for the creation of processes and their algorithms, but it makes such methods and techniques a subject of study in their own right. We may say that this leads to a system of meta-levels that is characterized through its *fractal self-similarity*. The structures of procedures for the development of processes, programs for the development of software, and languages for the design of *programming* languages all remind us of the matryoshka dolls seen on the book cover, as do recursive function definitions and the informatics-specific concept of *bootstrapping*.

Conceptually, Advanced Informatics rests on three pillars, which provide robustness and stability on all meta-levels because of the fractal self-similarity of informatics-specific structures:

Structure: Key for a good engineering design is the choice of a structure that meets the desired aims and funtionalities adequately, and that is pretty robust under demands of change management, the needed system scalability, and the required quality of service levels. The standard design principle here is that of an *Inductive Definition*, the mother of modular conception and design. Dependent on the construction at hand, an inductive definition has a number of advantages that apply to all meta-levels of consideration: it provides *concise and formal precision*, and gives us *compositionality* and also *well-foundedness*. Compositionality is the foundation for modular design, whereas well-foundedness is the key to inductive proofs that show correctness of the design. This approach to system design is at the heart of all logics and programming languages used in Informatics, and a good Informatician will deviate from this approach only in *very* well-reasoned and -justified circumstances.

Invariance: *If I had a fulcrum and a lever long enough, I could move the Earth.* This sentence of Archimedes was originally meant to refer to the application of physical force, but we may interpret it much more widely: in order to understand change, it is of central importance that we have knowledge of those things that do not change, which are therefore *invariant*. This concept pervades all of Informatics in the form of while-statements and recursion invariants, inductive hypotheses, homomorphisms of structure, fixed points in computational processes, the solutions of so-called constraint systems, and so forth. In fact, we make this widened interpretation of the above *Archimedean Point* a leitmotiv of all volumes of this trilogy.

Abstraction: As is the case in Mathematics, Informatics rests on the discovery and application of a suitable abstraction, whose design should follow with rigor the criteria for structure (preservation) and invariance. In fact, in Informatics we do not only have to master the design of a sole abstraction, rather the interplay of several layers of abstraction. And we need to understand what aspects of *reality* each such layer captures and represents. A suitable interpretation of the *Archimedean Point* here is that of the preservation of structure (known as *homomorphism*). This allows us to design and manage complex abstrations by decomposing them into simpler ones so that we can better understand and manage such complexity. One such homomorphism that is now an integral part of Informatics is that of *Type Analysis*. This allows for the (abstract) *execution* of real programs at compile time in order to investigate whether the program uses types correctly. This is a good example of how any *simulation environment* may be ideally constructed, so that its construction provides a solid basis for sound qualitiative statements.

The main objective of this trilogy is to make these three pillars graspable, in their essence, in their underlying pragmatics, and in their significance for Informatics as a field of study. To realize this objective, we naturally have to first cover a minimum of mathematical background, as this will enable us to conduct a deeper discussion of these central aspects of Advanced Informatics. In effect, Mathematics and Informatics are so tightly coupled that it is sometimes not even possible to draw a boundary between them. Such attempts at separation are especially difficult in Theoretical Informatics, with subjects such as Complexity Theory, Algorithms, and Automata Theory, as well as a range of subjects in Discrete Mathematics, Constructive Logics, Graph and Category Theory, and Numerics.

It is the very foundations of Informatics that are often characterized through a strong mathematical core. Many conceptual approaches, including but not limited to those found in Theoretical Informatics, have been borrowed from Mathematics. But there is also knowledge transfer in the other direction: there are classical problems in Mathematics that can be solved only with the aid of techniques from Informatics. A prominent example is the so-called *Four Color Problem*, whose solution rests to this day on auxiliary tools provided by Informatics. Kenneth Appel and Wolfgang Haken managed to reduce a positive answer to that problem to the investigation of 1,936 critical cases, which were then fully analyzed and confirmed with computer support. A computer here serves as an *instrument of scalability*, since it enables the treatment of relatively simple problems which are beyond the solution limits of humans for mere quantitative reasons: going through the 1,936 cases required 1,200 hours of computation time at that time.

Another approach that is characteristic of Informatics is rooted in the explicit separation of *syntax* (read: representation) from *semantics* (read: meaning) as developed in Linguistics. This separation gives us an additional and important degree of freedom in the design of bespoke structures with significant practical impact. Examples are bitvectors for representing subsets, matrices for representing homomorphisms of spaces, or (finite) automata for representation (formal) languages. We will see that this separation also allows for a new and fruitful perspective on classical mathematical problems and their solutions. In fact, playing with different repesentations may sometimes be the key to conceptual breakthroughs (cf. discussion of Kleene's Theorem in Chapter 6).

The purpose of this trilogy is not at all the communication and comprehension of individual mathematical results. Rather, it is intended to convey a sense for clear structures of computational thinking that have matured over the course of centuries. Mathematics, it has to be said, has without doubt produced the most mature domain-specific models. And those are jewels that Informaticians will do well to appreciate and to use for orienting their approach to a new application domain. The third volume is going to illustrate this concretely in its exposition of Linear Algebra, by demonstrating that we can move from the initial problem specification, to requirements derived from that, and to developed solution concepts in a seemingly effortless manner. This is facilitated by the elegant interaction of matrices and the more abstract linear maps that they represent. And it makes unmistakably clear that Mathematicians are also masters of this syntax/semantics design space. But one could say that Mathematicians see such designs as a means to an end, whereas an Informatician considers such interactions on meta-levels as a serious object of research and explicit study.

In Informatics, *meta-levels* are characterized through an interaction of structure, invariance, and abstraction. The importance of this concept is a direct consequence of its intent. Whereas in Mathematics we see the study of relatively few *natural scenarios*, and some of these scenarios (for example Number Theory) are subject to deep analysis for entire centuries, Informatics is a fleeting and short-lived activity in comparison. In Informatics, projects demand the constant development of new, often very specific, domain models and scenarios, whose half-life period may be very short. We might say that, in Informatics, *time to market* is much more important than *perfect product*. It is a central concern of Informatics to allow for the creation of conceptual and robust solutions even in the face of such enormous *time* pressures, by the provision of clear design principles and *meta-results* that facilitate reuse of solutions in new applications.

We therefore accompany the development of mathematical content in this trilogy with a systematic discussion of good style and its corresponding design criteria and rules. In the long-term, it is hoped to put the reader in a position in which she can clearly distinguish between *ad hoc* approaches and those that are based on conceptual design. In the latter ones, we make decisions that are problem-oriented, transparent, and goal-oriented – and we are aware of taking such an approach. Mathematical structures, results, and solution patterns serve as guidelines for adequate design decisions and for an acute awareness of long-term consequences that such decisions may bring about. The latter is incredibly important in practice, where in the final analysis the so-called *Total Cost of Ownership* is far more relevant than the cost of initial production.

Mathematics is traditionally a genuine hurdle for students of Informatics. Students often delay the taking of mathematics modules, if possible. And they often do not attend many lectures on mathematical topics. The reasons for this behavior may vary. But one reason seems to be that students of Informatics see Mathematics as a necessary and essentially superfluous evil. It is much more exciting to learn about and do system development, graphics, games, social networks, and other subjects that seem to be very remote from the concerns and toolkit of Mathematics. What students may not realize, and cannot be expected to know at the outset of their studies, is that all those subjects are permeated with Mathematics. Impressive advances in computer graphics (for example the skin reflections on the faces of the alien population in the movie *Avatar*) are not possible without deep mathematical analysis. By the same token, games and networks require an understanding of logic, discrete mathematics, and probabilities. Even software and system development is based on non-trivial mathematical foundations.

If we were to put this case to students, they might dismiss it by pointing out that "*I don't want to study Mathematics, I just want to use it*". Although this may partly be a justified response, we should keep in mind that even the mere *adequate* application of mathematical methods does require a foundational understanding of mathematical thinking. Otherwise, we may step into the trap of the "one-eyed practitioner" who thinks that *"Everything looks like a nail"* just because he happens to have a hammer. To be frank, the adequate application of Mathematics in Informatics is anything but simple. To determine – in a subject that is as abstract as Informatics – whether it is best to use a hammer, some pliers, or a screwdriver requires not only an understanding of the existence of such tools, but also of their deployment profile and caveats.

One purpose of this trilogy is to highlight the necessity of this core competency, which identifies the *tools*, their intent, their structure, and other properties so that we can show when and where use of such tools is adequate. This competency is not so much about the acquisition of a complete list of mathematical tools, as it is about the exemplary exposition of *Best Practice*. The foundational principle of *adequate* approaches requires a deep understanding so that it can be transferred successfully to other scenarios, perhaps with other tools, and even to future situations that we cannot begin to imagine at present. The foundations of Advanced Informatics are timeless in that sense, and represent a competency level of differentiating Archimedean Points of Informatics.[3]

The late ACM Turing Award Winner Robin Milner once responded to a remark *"The talk was not really that spectacular, I was able to anticipate all steps taken"* made by someone next to him in an auditorium with a question: *"Who is the better player of golf? He who drives the ball violently, and so time and again has to rescue his game out of precarious situations such as the deep bunker or the pond? Or he who continuously keeps the ball on the fairway?"* That's all that Milner said. In fact, it is not easy to answer this question, it does depend on one's point of view. The media and some spectators are bound to love the more spectacular playing style, even if its player will not win the trophy. In a similar way, we may celebrate a Mathematician who has found a proof for an important open problem but where this proof is only accessible to a small elite. An Informatician who has responsibility over systems on whose reliability thousands or even millions of people may depend, ought to keep the ball on the fairway. We may say that Advanced Informatics is the art of the control of the fairway, whereas Mathematics and Theoretical Informatics systematically practice plays out of the deep bunker. The reader may therefore ask herself repeatedly, upon reading and working with this trilogy, whether we are still on the fairway or – if not – where, when, and why we have abandoned it.

Approach of This Trilogy

Each of the three volumes of this trilogy has its own educational aims, which are summarized on page xx. Methods, principles, and results that are introduced in a volume are also picked up again and deepened in subsequent volumes.

---

[3] This trilogy is therefore not a substitute for existing introductory texts in Mathematics. Rather, its aim is to enrich such texts with the perspective of "Advanced Informatics".

Inductive Approaches is the first volume and establishes the ground rules for a powerful approach to the modeling and analysis of systems and specific domains. After a brief introduction to the required elementary mathematical structures, this volume focuses on the separation between syntax (representation) and semantics (meaning); and on the advantages of the consistent and persistent use of inductive definitions. In doing so, we identify *compositionality* as a feature that not only acts as a foundation for algebraic proofs familiar from school – which replace equals with equals – but also as a key for more general *scalability* of modeling and analysis.

A core principle that accompanies the development of this approach is that of an *invariance*, which we will discuss in different facets:

- as unchanging patterns in inductive definitions,
- as characteristic properties of inductive assumptions,
- as independence from choices of specific representations, and
- as structural preservation of functions or transformations (homomorphisms).

The intent here is to convey that invariance is a key for the mastery of change, be that change in the form of extensions, transformations, or abstractions. Such mastery comes with the search for and discovery of the respective *Archimedean Point*, which captures the core of a deeper understanding that brings with it a secure way of managing change: the more invariances we discover, the better we will master and control change! In the practice of software engineering, this insight is reflected in the *best practice* of conducting larger developments and adjustments in smaller, controllable steps.

Algebraic Thinking is the second volume and deepens the approach familiar from school – in which we *replace equals with equals* – to a general, axiomatic approach for mastering abstract structures and their properties. In doing so, we will examine *lattices* as a central structure in the processing of information, as well as classical algebraic structures such as *groups*, *rings*, *fields*, and *vector spaces*.

Each of these structures will be presented and analyzed in the same way through the following, fundamental criteria:

- axiomatic definition,
- sub-structures,
- factor structures,
- homomorphisms, and
- principles of compositionality.

As we do so, it will become particularly clear how successive structural extensions affect the above criteria. The latter is especially interesting for Informaticians, as it illustrates in an abstract manner the so-called *feature interaction problem*, a dreaded phenomenon in which the extension of system functionality (in this algebraic setting through the requirement of additional structural properties) leads to the loss of desired system properties. In this more abstract setting, we can observe how the very important concept of abstraction via homomorphism, which can be regarded as one of the most effective tools for any kind of mathematical modeling and, in particular, for so-called formal-methods-based system development, gets lost in the course of adding requirements: as witnessed by the *Fundamental Theorem on Homomorphisms*, homomorphisms introduce a well-understood theory of abstraction for groups and rings. Just adding one more structural requirement, the so-called absence of *zero divisors* (a part of the definition of fields) forces homomorphisms to be injective and therefore to be inadequate to impose abstraction. In addition, it also prohibits that fields are closed under Cartesian products. These shortcomings of fields lead ultimately to the introduction of vector spaces, which are closed under Cartesian products and have a very elegant theory of abstraction.

Of interest here is also the relationship between lattices and classical algebras: the sub-structures of an algebraic structure form a lattice! Such structural observations and insights are of general importance in Informatics since they help us to understand complex systems through a modular, *divide-and-conquer* approach of modeling and analysis.

Perfect Modeling is the third volume and introduces Linear Algebra – a subject matured over centuries and well familiar from school – from the perspective of a modeler of an application domain. The latter is a person with the task of making a new application domain both accessible and controllable. In discharging this task, the modeler is confronted with concrete and practically relevant problem statements for the given domain, which for Linear Algebra include:

- The solving of systems of linear equations, whose multitude of applications is already reflected in mathematics books used in schools.
- The description of the dynamics of discrete systems, including the understanding of population growth, the determination of stable states, the reachability of specific scenarios, and the reversibility of a course of action.
- The geometric construction and analysis of objects in higher-dimensional spaces, including concepts such as orthogonality, surfaces, volumes, the determination of cut surfaces, and reflections.
- Feature-based classification through machine learning, for example for recognition of handwriting or for recognizing faces through so-called eigenfaces (cf. page 11).

In hindsight, we can say that this third volume will have developed linear algebra in such a manner that we stay on a most comfortable *fairway*, on which solutions to initial and fundamental problems seem to appear effortlessly. In fact, we may interpret linear algebra with its numerous facets as the result of ideally executed *domain modeling*. The latter establishes a tailored formal language at a suitable level of abstraction, and offers a multitude of deduced concepts and functionalities – such as diverse forms of products, determinants, and decomposition principles, as well as methods for shift of representation, for the solution of systems of equations, and for the modeling of discrete dynamical systems.

This goal-oriented approach to the incorporation and systematic acquisition of new application domains is meant to produce mathematical content for Informaticians that illustrates, makes tangible, practices, and stresses the importance of a central mode of operation for Informaticians.

In the practice of the rapidly changing field of Informatics, it is nearly impossible to study new application domains with an approach that is even approximately as systematic and rigorous as the one we exhibit here for Linear Algebra. Nonetheless, this third volume is meant to provide competencies in individual structural techniques, but also a comprehensive sense of a deeper *domain comprehension*. And this will lead to an increased robustness when dealing with the interplay of conceptual complexity, key requirements, and appropriate pragmatics - an interplay that is typical for Advanced Informatics and its use in today's practice.

Meta-aims and Competencies of This Trilogy

The foundations of Advanced Informatics are primarily characterized through a series of meta-aims and their corresponding competencies. The latter will be acquired and deepened in a consistent and step-wise manner throughout this trilogy, illustrated by means of various scenarios.

Confident Mathematical Comprehension:

- mathematical precision
- corresponding formalisms, as well as
- mathematical approach

Application of Reusable Patterns:

- patterns for descriptions
- patterns for structuring
- patterns for proofs
- patterns for algorithms

Principled Approach:

- What is the core of a particular problem?
- What are appropriate solution patterns?
- How can we identify and apply appropriate patterns to solve a specific problem?
- How do we obtain a specific solution scenario (domain modeling)?

Mastery of Design Spaces:

- Separation of syntax and semantics: *How* versus *What*
- (Inductive) structuring
- Generalization and abstraction
  ... as foundations for principles of "divide-and-conquer" such as:
    - Invariance
    - Compositionality
  ... with the aims of gaining:
    - Correctness (for example inductive proofs)

      – Efficiency
      – Scalability

This first volume, *Inductive Approaches*, intones the leitmotiv of this trilogy *Mathematical Foundations of Advanced Informatics* through an interplay of structure, invariance, and abstraction. The two subsequent volumes, *Algebraic Thinking* and *Perfect Modeling*, reconnect with those themes by deepening their understanding in varying and ever more complex relationships. Our approach is therefore not too different from that of the French composer Maurice Ravel when he wrote the score for *Boléro*: the same melody is being played by an increasing number of musicians and instruments so that the listener can take in the leitmotiv, easily recognize it even in unusual circumstances, and transfer it to novel scenarios.

    The guiding aim of this trilogy is therefore the *competency of conceptual technology transfer*, which is not only critical to the success of concrete application projects, but which also features in job interviews at prominent ICT companies such as Google and Apple.[4]

Contract with the Reader

The motivation for writing this trilogy *Mathematical Foundations of Advanced Informatics* is rooted in the hope that the difficulties of mathematics can be overcome once the importance of the content of this trilogy has been understood. At that point, your engagement with the material will become more serious and dedicated. You will experience a first sense of achievement, and your studies with this trilogy will begin to become more fun. To support your personal engagement with the subject, this trilogy is rife with forward and cross references, excursions, and short thought experiments – in order to stimulate your own curiosity and research.

    In the age of the Internet and Wikipedia, it is easy to find information that supports your research, and the more you use such media the quicker and more effective you become at finding such information. We assume that you are already well familiar with such tools, and have experienced that some searches can dig up information that is completely unexpected and captivating at the same time. For example, you may find performances of Ravel's *Boléro* on YouTube.

    This trilogy utilizes these sheer unlimited possibilities of knowledge acquisition through online resources, and does this through the insertion of numerous forward references, whose deeper understanding requires additional knowledge not found directly in this trilogy. Such additional knowledge is gained over time throughout your degree studies or through continuous self-study that makes judicious use of online material. It is up to you to decide when, where, whether, and to what extent you would like to pursue the suggestions for further study made in this trilogy. In fact, experience seems to indicate that such suggestions bear fruit later on – even

---

[4] An excellent overview of the sorts of questions that applicants may have to be prepared for is given in [6].

if you may only have perceived them superficially at first, as in *"Hang on, didn't I read a remark about ... ?"*

We may compare your engagement with this trilogy in some ways with the manner in which you would learn a new language such as French, Chinese, or Arabic: best practice is that the entire learning will be done in the new language only, from day one onward, and where comprehension of the spoken language and the ability to speak it manifest themselves in a manner that can almost not be explained.

But this approach can easily lead to a feeling of "zero gravity": you will confidently discuss things that are highly complex, but you may be unaware of the fact that you may have lost the "road grip" needed for implementing these things in real applications. In that context, the exercises are an important control mechanism that checks whether you can transfer the foundations you have mastered to the solution of real problems. All exercises can be solved with the means and methods introduced in this trilogy.

Another central aim of this trilogy is to encourage you, the reader, to question things, to have fun with digging deeper, with rummaging through other sources, and with trying things out for yourself.[5]

The fact that you can indeed just try it yourself and how you might go about this are illustrated in the further reflections made throughout this trilogy, for example in the reflections on inductive principles based purely on grammars on page 150 of this first volume.

We hope that this trilogy will also help you with initiating and conducting dialogues with your peers, teachers, or others about the questions you raise and insights you gain from working with this trilogy. You should note that such communication and the feedback that comes from it are important for the recognition of your work, for your sense of enjoyment of this work, and for your motivation to work more on that material. The behavioral economist Dan Ariely captured the essence of this beautifully in his *TED Talk* available at http://www.ted.com/talks/dan_ariely_what_makes_us_feel_good_about_our_work.

When engaging in such dialogues, you may serve in more than one role of genuine value to you and others: you may primarily be the student who needs to become motivated to master and excel in this mathematical subject; but you may equally be the student who will infect his or her peers with his or her sense of fun and achievement so that your peers will also enjoy and master this material together with you.

Limits of Online Resources: As mentioned above, online resources such as Wikipedia are an ideal means for obtaining stimulation, retrieving basic or background information about concepts, and clarifying formal notation,[6] in order to

---

[5] For readers of the German language we recommend the book "Mathematik für Informatiker" by Gerald and Susanne Teschl [78] as especially valuable accompanying reading. Interesting English textbooks on the topic are those of John Vince [85] and Gerard O'Regan [61].

[6] There is generally no binding standard for notation. Depending on context or social customs, different notations have been adopted, often with overlapping usage. It is therefore part of a mathematical training to gain a certain interpretative security in the study and usage of notation. Searching online references is an ideal means of gaining such security.

build up a solid *general knowledge* of a specific subject. Based on its global availability, enormous diversity, and astonishing topicality, Wikipedia is far superior to text books and lexicons. But you should take great care in not misunderstanding this superiority: *Wikipedia and many other online resources are not scientific sources that can be cited in scientific writings or that can be highly trusted for accuracy and completeness.* The reason for this is that the information provided in those resources is generally not subjected to *peer review* of scientific experts in the domain pertaining to that information, that the sources of those pages can be modified in uncontrollable ways, and that proper citations should point to immutable sources.[7] It is fortunate, though, that Wikipedia pages on subjects in Mathematics and Informatics are generally of very high quality, regardless of the fact that those pages may be edited over time.

Hosted Web Site: We have set up a Web site for additional material under the URL `http://www.higher-informatics.org`, where particularly model solutions for exercises can be found. It should be noted that each chapter of this volume is supplemented by numerous exercises which have two principle directions: exercises addressing simple issues purposely left open for the reader and exercises amplifying the reader's understanding of the material. Moreover, we will try to manage an up to date errata list on this web site in order to resolve known issues with the printed edition.

Notational Convention

As mentioned above, this trilogy provides numerous excursions, forward references, and illustrative discussions in order to further motivate and deepen its material, and to provide an impression of the extended applicability of the material acquired by the reader. In order to not disrupt the flow of reading, especially on first reading, this supplementary material is given a grey background or is grouped together in a section called "Reflections" at the end of the respective chapter.

Acknowledgments

The idea for the envisioned trilogy is based on a history of over twenty years of teaching at the University of Passau, the Technical University of Dortmund, Kansas State University, and Imperial College London. It is therefore informed by many discussions with former colleagues, especially Olaf Burkart, Alfons Geser, Hardi Hungar, Jens Knoop, Gerald Lüttgen, Markus Müller-Olm, and Michael Mendler, and strongly influenced by the experiences gained in numerous industrial projects

---

[7] *Digital curation* is a recent trend and research area in Informatics that aims to combine the advantages of online resources with the permanency of conventional sources such as textbooks and research monographs.

with partners such as BASF, IKEA, Siemens, ThyssenKrupp, and T-Systems. The idea for this book project was formed after a decision taken at the University of Dortmund to move the mathematical training of Informatics students into its own department. The implementation of this transition would not have been possible without the initial support of Gabriele Kern-Isberner and Hubert Wagner.

This book can be regarded as an enhanced English version of its German counterpart, which was co-authored by Malte Isberner who left after his Ph.D. for Google, Mountain View, at the end of 2015. Malte significantly influenced the German book. Moreover, we are very grateful to Alnis Murtovi and Maximilian Schlüter for carefully proofreading the manuscript and their support concerning the model solutions for the excercises. Many other people kindly supported us in the form of proofreading and recommendations. We cannot name them all here, but we would especially like to mention and expressly thank Stefan Naujokat, who provided the images for the Towers of Hanoi, and Stephan Windmüller and Nadine Neumann, who provided the image of the telephone book. Additional support, suggestions, and valuable feedback came from Oliver Bauer, Radu Grosu, Anna-Lena Lamprecht, Maik Merten, Heinrich Müller, and Johannes Neubauer. The most frequent discussion partner through all those years that saw the conception and production of this first volume was Tiziana Margaria, an engineer whose practical outlook markedly informed the conception of this book and research done together with the team at the TU Dortmund. Finally, we would like to thank the staff at our publisher Springer, especially Mr. Hermann Engesser, Mr. Ronan Nugent, and Mr. Alfred Hofmann, whose support made this book project possible.

Dortmund, April 2017   *Bernhard Steffen*
                                                                                    Oliver Rüthing

London, April 2017   *Michael Huth*

Contents

Chapter 1
Introduction

> *There is no silver bullet for problem solving in Advanced Informatics.*
>
> *(Based loosely on Euclid)*

There is no doubt that Mathematics, next to Electronics and Electrical Engineering, is one of the founding pillars of Informatics. Historically, Informatics might even be seen as a particular aspect of Applied Mathematics. For example, as relatively recently as the 1990s, RWTH Aachen in Germany had a Professorial Chair entitled *Praktische Mathematik, insbesondere Informatik* which translates to *Applied Mathematics, in particular Informatics*. There is no doubt, though, that Informatics has come of age and has grown into its own scientific and engineering discipline that trains its own professionals. But this does not mean that the mathematical roots of that discipline have become less relevant over time, quite the contrary. Those foundations of Informatics that are particularly well understood and established are typically characterized by a solid mathematical core. Also, many patterns of conceptual approaches in Informatics, including those found in its theoretical branch, are borrowed from Mathematics. What distinguishes Informatics from many other disciplines profiting from mathematical results is its way of pay back: in Mathematics, be it Applied or Pure Mathematics, there are problems, for example combinatorial ones, whose solutions can only be achieved with tools from Informatics.

One well-known such combinatorial problem is the so-called Four Color Problem:[1] given a planar map of countries and their borders, for example the new map of Europe drawn up at the Vienna Congress in 1815, can we color each country with a color such that

1. no two countries that share a border have the same color
2. no more than four colors are used to color the entire map?

In 1976, Kenneth Appel and Wolfgang Haken derived a proof of this claim as follows. First, they developed a suitable notion of *size* of a map, which allows us to compare any two maps in terms of their sizes. Since a size is a positive and integral number such as 7, it follows that if there is a counterexample to the above claim, then there is a counterexample M_{min} of *minimal size*. Now, Appel & Haken identified 1,936 different maps $M_1, \ldots, M_{1,936}$ with the following properties:

---

[1] See http://en.wikipedia.org/wiki/Four_color_theorem

© Springer International Publishing AG, part of Springer Nature 2018
B. Steffen et al., *Mathematical Foundations of Advanced Informatics*,
https://doi.org/10.1007/978-3-319-68397-3_1

N None of these maps $M_1, \ldots, M_{1,936}$ can be a submap of a map that is a minimal counterexample.

E Each map M that cannot be colored with four or fewer colors must contain one of the maps $M_1, \ldots, M_{1,936}$ as a submap.

Note that this argument appeals to submaps as *substructures*, a topic we will explore in the second volume of this trilogy. Appel & Haken could then appeal to a proof principle that we will study in this first volume: if there is a counterexample, i.e., a map that cannot be colored with four or fewer colors, then there must be such a map M_{min} of minimal size, which, according to item E, is additionally guaranteed to have some M_i from $M_1, \ldots, M_{1,936}$ as a submap. This, however, directly contradicts item N above, which proves the Four Color Theorem by contradiction. The proof principle of contradiction will be one of the typical proof patterns which we will study in this first volume.

The details of this argument required the inspection of too many cases to have them verified by humans in a reasonable amount of time and with enough diligence. These cases were therefore analyzed with computer programs, making a computer an *instrument of scalability* for this mathematical proof. This approach to mathematical proof was somewhat revolutionary and highly contentious. Some mathematicians objected to the fact that a computer program was used in a mathematical proof, and so the program and its execution environment would have to be trusted to correctly capture the case analysis required for completing that proof. Some said that proofs constructed by humans, typically written in a semi-formal style, are to be preferred since they can be checked more reliably by mathematicians – but the sheer scale of the argument needed for the Four Color Theorem prevented such an approach.

There is a very deep theorem in Theoretical Informatics on *Probabilistically Checkable Proofs* that says that proofs of a certain expressiveness can be verified with high probability by first transforming them into a specific form and then just inspecting the correctness of some random parts of the proof in that transformed form. This is an example of an important principle in Informatics, the transformation of one representation into another one in order to simplify a problem without actually changing its meaning. This trilogy will study such transformation principles already in the first volume.

Concepts and tools developed in Informatics have shaped modern ICT infrastructure such as the Internet. And this infrastructure can, in turn, be used to solve open problems in Mathematics and Informatics by sharing computing and human resources, for example through *crowdsourcing*. In principle, one could use that approach to complete the case analysis needed for the Four Color Theorem.

This approach was actually used successfully (in that mathematicians generally accepted the resulting proof) on the Hales-Jewett Theorem[2]: Imagine a cube of length n to be divided into $n \times n \times n$ subcubes, called cells. We can do that same division for *hypercubes*, cubes in dimensions $k > 3$ (then we have n^k cells). Now suppose that we color each cell of such a cube with one of $c > 1$ colors – where the

---

[2] See http://en.wikipedia.org/wiki/Hales-Jewett_theorem

value of c is fixed. The theorem then says that in sufficiently high dimensions (i.e., for a sufficiently large value of k) any such coloring needs to contain a row, column, or diagonal of n cells that all have the same color. This means that higher-order versions of the game of Tic-Tac-Toe[3], which we study in this first volume, have very interesting strategies that cannot exist in dimension $k = 3$. Alas, we do not seem to be able to describe such strategies in a manner in which we could execute them. This illustrates that, in Informatics (like in Mathematics), there is sometimes a gap between proofs of existence and the ability to use such a guaranteed existence in a practical way.

We conclude from the above discussion that Mathematics and Informatics are intimately connected and often co-dependent. At times, it is not even possible to say whether a problem or its solution is in essence a part of Mathematics or Informatics. It is particularly difficult to make such a distinction in the area of Theoretical Informatics, with research topics such as Complexity Theory, Algorithms, Automata Theory, Discrete Mathematics, Constructive Logics, Graph Theory, Category Theory, Numerics, and so forth. There is little doubt, though, that much of Informatics today would be inconceivable without the use of tools and insights from Mathematics. Conversely, Informatics and its evolution exercise a sustained and marked influence on Mathematics as it is practiced today.

Advanced Informatics

Advanced Informatics is a result of the aforementioned interconnection and co-dependency of Mathematics and Informatics; its aim is to develop tools for the successful transfer or adaptation of mathematical results and approaches to problems encountered in Informatics. This trilogy is therefore not primarily concerned with the dissemination and explanation of mathematical results. Rather, it is intended to convey patterns of thought that matured over the course of centuries and explore how these patterns can be used to solve problems in Informatics. We may say without any contention that Mathematics itself offers some of the best-developed models of application domains – be it for Physics, Chemistry, Engineering, Social Sciences, or other areas of inquiry. It is thus reasonable to say that informaticians stand to benefit if they know such methodological gems and reflect on them whenever they study problems in a new application domain.

This trilogy will therefore not only develop mathematical concepts but also accompany such development with a discussion of good style and methodology and with an articulation of the corresponding design criteria. It is thus hoped that readers of this trilogy will gain the ability to clearly distinguish between approaches that are *ad hoc* and approaches that are founded on conceptual design principles, and then be able to make decisions that are conscious, problem-oriented, transparent, and goal-directed. To that end, mathematical structures, results, and methods will serve as guidelines for making adequate design decisions and for sharpening the aware-

---

[3] See http://en.wikipedia.org/wiki/Tic-tac-toe

ness of long-term consequences that such decisions may bring about, for example within the lifecyle of an ICT system. The third volume of this trilogy will illustrate this approach concretely through the example of linear algebra: an area in which the initial problem statements, the resulting requirements, and developed solutions are derived in a seemingly effortless way.

Additionally, the discussion of and emphasis on structure should help the reader to better understand the nature of Informatics, especially in comparison to its cousin Mathematics. In the final analysis, Informatics is broadly speaking also an engineering discipline, which means that it cannot only produce aesthetically pleasing and elegant insights and methods; it also needs to produce workable solutions in current and future ICT systems. It is interesting, though, to observe that recent insights from Theoretical Informatics can have considerable practical significance; let us mention work on logic-based software verification and the fact that such work has created start-ups (see for example [60]) that, in some cases, have been bought by global ICT leaders with strategic intent. This also illustrates that the pursuit of simplicity and elegance, for example in a logical formalism, may well have huge practical benefits: a clear and clean approach and a design that concentrates on the essence of a problem are an investment that often pays off in the long-term – and Advanced Informatics provides us with the tools for making such investments.

Below, we will discuss some scenarios with typical problems encountered in Informatics in Section 1.1; tools for solving such problems will be developed in this trilogy. Then, we present a fundamental principle of Informatics, the explicit separation of syntax and semantics, in Section 1.2. In Section 1.3, we provide some personal reflections on the journey that you will undertake when reading and working through this trilogy. In Section 1.4, this first chapter then ends with an overview of the learning outcomes and skills that this first volume should help you realize.

1.1 Illustrative Examples and Typical Problems

Euclid's Algorithm for Computing the Greatest Common Divisor

The *Greatest Common Divisor (gcd)* of two natural numbers a and b is the greatest of all natural numbers that divides a as well as b. For example, 5 is the *gcd* of 50 and 65. Algorithms for automatically computing gcds have many applications; for example, they are a key component in the modern *public-key* encryption system called the RSA algorithm[4] [65]. It is quite remarkable that algorithms for the computation of *gcd* used in today's practice are derived from a method that is more than two thousand years old, the so-called Euclidean Algorithm[5].

We may express this algorithm informally as follows: we subtract from the larger number a a maximal multiple of the smaller number b until b becomes zero. The *gcd*

---

[4] See http://en.wikipedia.org/wiki/RSA_(cryptosystem)
[5] See http://en.wikipedia.org/wiki/Euclidean_algorithm

of the original values of a and b is then the current value of a. This requires some bureaucracy: what to do when a is no longer the larger number and – related to that – whether the case when a equals b requires special attention. The algorithm also requires an understanding of what *maximal* multiple means here: updates should preserve an *Invariant*, that a and b are natural numbers. Figure 1.1 illustrates this algorithm with $a = 65$ and $b = 50$: the first update gives us new values $a = 50$ and $b = 15$ (ensuring that a is larger); the next value pair is $a = 15$ and $b = 5$. Finally, this results in $a = 5$ and $b = 0$, when the algorithm stops with $a = 5$ as the greatest common divisor of 65 and 50.

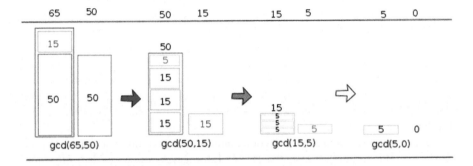

Fig. 1.1 Illustration of the Euclidean Algorithm with input $a = 65$ and $b = 50$

A first, superficial inspection of this method does not at all make clear whether the Euclidean Algorithm correctly computes the *gcd* of two natural numbers. But such clarity arrives as soon as we realize a crucial *invariant*: that after each step the *gcd* of the newly computed value pair is the same as the *gcd* of the currently given value pair (see Theorem 2.1 on page 33)!

Invariants are a central key for unlocking the comprehension of iterative programming constructs and recursive computations. This insight is in complete agreement with the intuition that goes back to Archimedes:

If I had a fulcrum and a lever long enough, I could move the Earth.

The interpretation of invariants as Archimedean Points[6] is a good guideline that we will follow in this book: it helps us to appreciate and master the reasoning about iterative and recursive programs; it allows us to do inductive proofs over sets that are iteratively defined or whose objects are structured in tree-like form; and it enables a sound approach to evolutionary program development and change management [73]. All this is possible by asking, and successfully answering, *What is the constant in all this change?*

---

[6] See http://en.wikipedia.org/wiki/Archimedean_point

The Towers of Hanoi

Consider the following task: n discs of decreasing size need to be moved from a source rod to a target rod by using an auxiliary rod, but where only one disc may be moved at a time, and no disc may be placed on a disc smaller than itself. The question we would like to answer is the following:

Question: Is this task achievable? And if so, how can we solve this task in a minimal number of steps that move discs to another rod without breaking the rules?

This illustrates that many problems in Informatics first need an understanding of whether they can be solved in principle. Given that, we can then investigate how best to solve such problems and so we seek *optimal* solutions. For the Towers of Hanoi, it should be clear that the minimal number of steps is a good measure for what is "best". More complex problems, however, may have several, perhaps conflicting criteria for optimality; consider the task to implement a secure communication infrastructure for a network consisting of sensor nodes and some servers: encrypting data will improve security but will also consume precious energy on sensor nodes that run on batteries.

In the previous example of *gcd* computations, the algorithm was given. For the Towers of Hanoi, another essential part of Informatics is now needed: *algorithm design*. Figure 1.2 illustrates the inductive and elegant character of solutions whose descriptions lend themselves to an investigation of the required solution steps. The Archimedean Point (the Invariant) of the preceding example is used to show the *correctness* of the Euclidean Algorithm. The Archimedean Point for the Towers of Hanoi informs both the *design* of an algorithm and the reasoning about its *correctness* – as you will see in Chapter 4.

Retrieval of an Object from a Collection

Many algorithms will work better for some inputs and worse for others. The Euclidean Algorithm, for example, does not update the values of a and b the same number of times; those numbers depend on the input values of a and b. For such algorithms, it is important to understand their *worst-case* behavior over the specified input space. Let us consider an example:

How many objects from a collection of objects do we have to remove from that collection in the worst case to determine whether a specific object is in that collection?

As for the Towers of Hanoi, we do not have an algorithm but first need to design one and then answer this question for it. This abstract situation is captured in Figure 1.3. It should be pretty clear that there is a trivial algorithm that determines this: keep removing objects from the collection until you find the specified object. This requires that the algorithm can decide whether two objects are equal (to ensure correctness), and that there are only finitely many objects in the collection (to ensure termination).

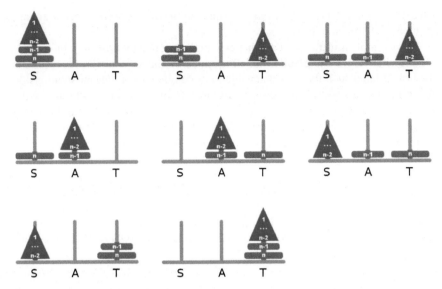

Fig. 1.2 Towers of Hanoi: an inductive description

However, in the worst case this algorithm will have to remove *all* objects from the collection. This happens in two kinds of scenarios: those in which the object is not in the collection, and those in which it is in the collection only once and all other objects are removed first.

Fig. 1.3 Searching for an object in a collection: mathematical objects in a bucket (left) and phone numbers in a phone book (right)

Can we design algorithms that behave better than that?

In other words, are there solutions to this problem whose worst case is always better than removing all elements from the collection? The answer to this depends on the *rules of the game*. Are we allowed to determine the structure of the collection ourselves, i.e., may we determine the manner in which objects are organized into a collection? If so, then we can do much better. Imagine a phone book and think about why it is possible that we can look up numbers in that book so quickly amongst thousands of similar entries.

This example illustrates yet another essential aspect that is typical of Informatics: the exploitation of a **design and modeling space**. We may (within limits) *re*structure the task at hand such that it is amenable to more efficient or elegant solutions. For the task at hand, the key to success is the *sorting* of a collection – be it based on the inherent order between objects (for example the lexicographical ordering [31] of names in a phone book) or on more abstract orders (for example those based on search keys). The ability to successfully exploit a design space in this manner is a central art that a working Informatician should master.

In the next section we will focus on principles and approaches for mathematically controlling systems with the aim of establishing their reliability and correctness. We will therefore not say much about methods for improving the efficiency of algorithms and of the corresponding representations of data that they manipulate; we refer the reader interested in these topics to the excellent textbook by Cormen et al. for further details [17].

Correctness of Programs and Its Demonstration

In Informatics, the correctness of algorithms is often separated into two aspects: so-called *partial correctness* (see also page 177) demands that all terminating computations of the algorithm render the correct results. This notion of correctness therefore makes no claim about the *termination behavior* of the algorithm. A drastic example that illustrates this separation well is the so-called Collatz function[7], whose *raison d'être* seems to be to compute the result 1, independent of its input value.

> **Input:** Positive natural number n
> **Output:** 1
> **while** $n \neq 1$ **do**
>   **if** n *is even* **then**
> | $n \leftarrow n/2$
>   **end**
>   **else**
> | $n \leftarrow 3n + 1$
>   **end**
> **end**
> Print out n.

Algorithm 1: Computing the Collatz function

It should be pretty obvious that the Collatz function is partially correct: whenever this function terminates at all, it is guaranteed to render the correct output 1 as any value different from 1 forces yet another iteration of the computation. At the time of writing, it is an open problem whether the Collatz function terminates for all positive natural numbers as input. You may be surprised by this, given that the

[7] See http://en.wikipedia.org/wiki/Collatz_conjecture

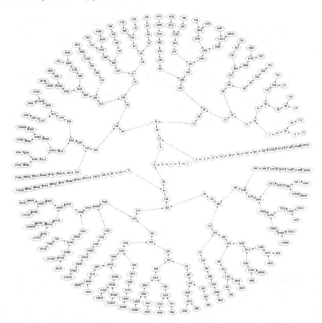

Fig. 1.4 Graph showing the dynamics of the Collatz Function for various inputs of positive natural numbers, up to a depth of 20 (Source: http://commons.wikimedia.org/wiki/File: Collatz-graph-20-iterations.svg)

iterative dynamics of this function are so easy to describe and grasp. But we can get a feeling for how unpredictable these dynamics may be when we take the input value 27; the value of n then becomes as large as $9,232$ before it converges down to 1. In contrast, for the input value 100 all subsequent values of n stay below 100 until it reaches 1.

The Collatz function determines a graph on the set of positive natural numbers, where we draw an edge from n to n' if either n is even and n' equals $n/2$, or if n is odd and n' equals $3n+1$. This graph therefore captures the dynamics of the Collatz function. Figure 1.4 shows this *Collatz Graph* for a depth of up to 20, suggesting that these dynamics do not follow a discernable pattern.

Partial correctness and termination of programs are typically proved with inductive approaches. That such approaches are unable to prove termination of the Collatz function indicates that this function strongly violates the design guidelines that we propagate in this first volume. We will explore and discuss this violation further in Section 6.2 of the final chapter in this book.

Management of Bounded Resources

In Informatics, it is common to distribute the solution of complex tasks to different problem solvers. Modern ICT infrastructures make it possible to utilize many, many

Fig. 1.5 Schematic concept of the *ANTS* Mission (Source: [80])

such solvers, using *crowdsourcing* techniques. One such example, rather modest in size, is the *ANTS* mission of the US space agency NASA, sketched in Figure 1.5. The task is that about two thousand spacecraft, each one the size of a standard notebook, have to navigate through an asteroid belt to complete a mission – for example, taking photographs, collecting samples, conducting chemical analyses, and so forth. Since the spacecraft are limited in size and so also have substantial resource constraints, the mission requires a clear and advantageous distribution of roles and coordination of activities amongst these 2,000 spacecraft. Each spacecraft fulfills a specific duty and its results need to be communicated and combined with those of other space-craft in a manner that supports the mission objectives. Generally speaking, it is very hard to coordinate such a complex cooperation efficiently and effectively. Therefore, we want to condense and abstract this discussion to convey very simple exemplary problems pertinent to this application.

Suppose we have a swarm of minicomputers, each one of them being able to compute arithmetic for integers up to one byte. These minicomputers can compute the usual arithmetic operations $+$, $*$, $-$, and $/$, where $/$ is only defined for divisions that have no remainder.

Questions: Is it possible to guarantee that all computed results that are representable with one byte are correct, despite the possibility of carry-over problems? Is it further possible to combine such results into overall results represented in four or eight bytes without any loss of precision – say by using a more powerful computer stationed on earth?

Our treatment of quotient structures in the second volume, *Algebraic Thinking*, will demonstrate that it suffices to perform such computations only on the remainders of divisions based on a suitable set of numbers that are pairwise relatively prime, i.e., that have 1 as their only common divisor. Our development of such algebraic abstractions will therefore show how and why the questions asked above have an affirmative answer. Quotient structures have many applications in Informatics, in particular in modeling and in automated program analysis; for the latter, quotient structures feature prominently in Type Theory and in the Theory of *Abstract Interpretation* [18].

Solving Systems of Linear Equations

In your Mathematics or Physics courses in school, you often had to understand how systems of linear equations adequately model problems, and how small such systems can be solved manually. Therefore, there is no need to motivate the utility of such systems and their solutions. But it is likely that your courses did not address a crucial aspect of this approach, its *scalability*. How do we solve a system of linear equations with many unknowns? Complex systems such as chemical engineering plants may have models with hundreds of thousands of unknowns, for example. And they may approximate non-linear chemical processes with linear equations.

Linear Algebra is a mathematical framework that not only provides efficient methods for the solution of systems of linear equations; it is also a *domain-specific* framework in the language of Informatics: it elegantly captures the structure of solutions and their potential uniqueness, and influences the transformations that may modify such systems of equations. The third volume of this trilogy develops and discusses the mathematical foundations of linear algebra from the perspective of a *Domain Modeler*, an individual whose task it is to design a new application domain (here "Linear Algebra", a domain tailored to solve linear equations) to be both accessible and controllable. The concepts of linear algebra are powerful enough that they are usable in all kinds of applications and situations. Let us mention here the feature-based classification techniques in *Machine Learning* and the method for face recognition sketched below.

Eigenfaces The central question here is how we may characterize individual faces in a maximally simple manner that allows the computer-aided recognition, production, and classification of human faces. A very successful approach to solving these recognition and classification problems is rooted in Linear Algebra: faces are approximated by a mathematical combination of information drawn from a set of representatives – the so-called *eigenfaces* [81]. For Informatics, this approach is especially attractive since it combines the idea of *approximation* with a method from Linear Algebra, the decomposition of the approximating space into *eigenspaces*. Approximation is a form of *abstraction*: the high-dimensional feature space of a real face is reduced in dimension by projecting this information onto eigenspaces of much lower dimension. The third volume of this trilogy will make eigenvectors and eigenfaces concrete topics of investigation. Further details on eigenfaces and their applications can be found in textbooks on specific research areas such as Computer Graphics.

 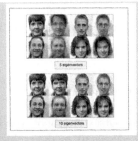 

Fig. 1.6 Eigenfaces (Source: [29])

The left picture shows a mean image (upper left corner) and 19 eigenfaces. Mean images are obtained by the superposition of all the underlying faces. They are therefore a good starting point for producing specific variants of faces through the introduction of eigenfaces of different intensity. Mean images are thus the source of the complete *Face Space*. Other images of faces are then obtained by a so-called linear combination of eigenfaces. The quality of these constructed images will depend on the chosen dimension of the corresponding approximating mathematical space. Figure 1.6 shows variants of this approach for approximating spaces of dimension 5, 10, 15, and 20 (respectively).

1.2 Domain-Specific Languages and the Power of Representation

It is traditional in Mathematics to argue precisely but to do so in an informal natural language such as English. In that practice, there is usually no formal separation between the language level in which arguments about a problem are conducted and the layer in which the problem itself is expressed. In particular, mathematicians tend not to make a conceptual separation between the so-called representations (*syntax*) and their intended meaning (*semantics*). For informaticians, making such a clear separation is however essential. This trilogy aims to demonstrate two advantages of such a conceptual separation:

- **The formal control of the representational level.** Such control is the foundation for any computer-aided processing, ranging from simple text editing and the translation of programs into machine-operable codes to ambitious tasks such as automated theorem proving, compiler generation, and computer-aided natural language processing.
- **The creative elbowroom that such a separation enables.** The intuitively given problem can, like in Physics, be seen as *naturally given*. The conceptual separation gives us flexibility in how the problem is formally described and represented. A related advantage of Informatics over the Natural Sciences is that we may *invent* formalisms that have no counterpart in the natural world but that solve real-world problems.

Example: Natural Numbers. Today we are used to the decimal representation of numbers, its logarithmic conciseness, and the elegant manner in which we can compute in that representation in reference to specific positions of its digits in an almost mechanical fashion. Historically, it took a very long time with numerous evolutionary steps until we arrived at this representation. Most basic was the unary system, where a number is represented by vertical bars that "count out" that number. This obviously has problems with scalability: representing large numbers is awkward. To overcome this, e.g., Roman numbers were invented, which soon turned out to be complicated to compute with, etc. The elegance of the place-value systems, which appeared as early as 1800 before Christ in Babylonian texts, was only recognized much later in Europe.

Already this simple example illustrates the mentioned advantages of the separation of representation and meaning: it allowed the development of an extremely concise representation of numbers that, at the same time, optimally supports computation without touching the intended meaning. The underlying principles, *logarithmic representation* and *compositionality*, are ubiquitous in modern "representation engineering", which is particularly popular when designing domain-specific languages (DSLs).

Foundational for systematic *representation engineering* are inductive approaches, the topic of the first volume of the trilogy. Rigorously applied they allow one to align the structure of the syntactic representation with the conceptual semantic structure, which is the key to modular design, compositionality, and the powerful proof principle of induction.

Inductive definitions, in particular in the popular Backus-Naur Form (BNF, see Section 4.3.3) are key for the design of (formal) representation languages such as programming languages, specification languages, logics, etc.

In fact, BNFs themselves form an inductively defined (formal) specification language: a language to inductively define languages. It is therefore customary to call a BNF a *meta-modeling language*, as BNFs are designed to specify/model the structure (syntax) of (modeling) languages. Remarkable here is that BNFs are *reflexive*, meaning that it is possible to describe the structure of BNFs using BNFs themselves. Powerful frameworks such as Eclipse's *Meta Object Facility* (MOF) generalize the meta-modeling idea to enable systematic construction of domain-specific languages, and even entire according development environments [52].

We will illustrate the power of BNFs with numerous examples. In particular, we will show how the five Peano Axioms can be reduced to just one line of specification when using BNFs. We will see that this is not just a matter of the syntax of BNFs but of its inherent semantics, i.e., of the way BNFs are generally interpreted (see Section 4.4). This emphasizes another dimension of good (domain-specific) language design: optimally, a language should allow one to exclusively focus on the intended purpose, and automatically take care of surrounding "common knowledge". In other words, perfect (domain-specific) languages factor everything out which is a priori given [53]. For the BNF-based specification of natural numbers these are, e.g., the syntactic notion of equality and the minimality requirement imposed by the fifth Peano Axiom (see Sec. 4.1.1). In the ideal case, DSLs allow one to focus on the primary purpose, the WHAT, and leave the corresponding technical realization, the HOW, to the DSL framework. Parser generation is a prime example for such a WHAT/HOW separation (see Section 4.3.3).

This concept of factoring out is inherent also in many representation formalisms that are not based on BNF. Examples are *Hasse diagrams*, which express partial orders relative to the assumed transitivity of the relations (see page 168), Binary Decision Diagrams (BDDs)[8] for representing Boolean functions, automata that specify potentially infinite languages with their imposed notion of acceptance, temporal logic formulas that characterize either languages, automata, or other kinds of systems relative to their notion of satisfaction, and many more.

One crucial point is always the finiteness of representations for potentially infinite artifacts, such as subsets of infinite sets, functions with infinite domains, computational trees, and so forth. This leads to the notion of expressivity: which of these potentially infinite artifacts can be described, and is the set of describable artifacts adequate for a certain scenario?

We will see that already very basic counting arguments suffice to prove that not even every function from the set of natural numbers to itself can be programmed. The point is that the set of computable functions is simply limited by the fact that a program, written in whatever language, constitutes a finite representation. However, there are only countably many finite representations, in contrast to functions with infinite domain, of which there are uncountably many (see Sec. 3.2.3).

An intuitive example of the described limitations are quantifiers in First-Order Logic (FOL) (see Sec. 2.1.2). FOL can be regarded as an extension of Propositional Logic (see Sec. 2.1) designed to capture a fragment of infinite formulas (in case that the underlying domain is infinite, as, e.g., for the natural numbers). It allows us e.g., to express that every natural number has a successor, a property which would have to be spelt out explicitly for every value otherwise. In this sense, the so-called universal quantification is nothing but a very specific infinite conjunction, whereas the existential quantification corresponds to such an infinite disjunction.

In fact, one could say that it is the choice of the adequate restrictions which provides domain-specific languages with their power. FOL is a perfect example for such an adequate choice, as are the various temporal logics which restrict quantification to capture a notion of time.

The general principle of finitely capturing infinite artifacts is a certain kind of regularity: there is a finite description mechanism that allows one to describe how to obtain/capture infinite artifacts. Typical such description mechanisms provide a finite kernel together with rules for extension or completion. Classical such constructions are given by the natural numbers that start just with one element but with the requirement that increasing a natural number by one is still a natural number (cf. Def. 4.1), the move to integers which are additionally closed under subtraction, rationals which are closed under division, and the reals which are closed under limit construction (see the second volume of this trilogy). Popular examples in computer science are, e.g., the unrolling of loops in programs or graphs, the Kleene hull

---

[8] See http://en.wikipedia.org/wiki/Binary_decision_diagram

(see page 137), the successive construction of new elements via BFN rules (see Sec. 4.3.3), the completion of lattices by adding suprema and infima (see the second volume of this trilogy), and the closure under certain laws, such as transitivity, reflexivity, and symmetry for equivalence relations (see Section 3.4.4). The resulting artifacts and concepts are often named differently in different contexts, such as, e.g., hulls, completions, extensions, closures, (linear) spans.

All these examples use the power of *implicit representation*, which allows one to work with the (infinite) artifacts without having to construct them in their entirety. The third volume of this trilogy, "Perfect Modeling", illustrates this principle by means of the traditional mathematical discipline of linear algebra, where matrices are used as a convenient means to reason about vector spaces. In computer science, this principle of implicit representation is not only used to describe infinite artifacts, but also to obtain concise representations of finite artifacts. Popular examples are the abovementioned Hasse diagrams to describe partial orders (see page 168), Binary Decision Diagrams (BDDs) to describe Boolean functions, bitvectors to describe power sets (see pages 69 and 92), and, more generally, any form of intensional characterization (see page 43).

Important in all these cases is that the representation does not only allow one to characterize infinite artifacts, but also to work with them. We will emphasize this aspect of the inductive approach in this book. In particular, we will see that BNFs, applied adequately, allow one to elegantly design domain-specific languages, i.e., languages that are concisely represented and tailored to serve specific purposes (see Sec. 6.3). Important here is that the syntactic structure defined by the BNF mirrors the intended semantic structure. This is fundamental for computer-aided processing, ranging from simple text editing and the translation of programs into machine-operable codes to ambitious tasks such as automated theorem proving, compiler generation, and computer-aided natural language processing.

The impact of representation can, however, reach much further. It underlies the adequate choice of data structures in programming, and it may even help to establish the mindset for solving open problems: for example, the detour via a representation in terms of automata was the key for solving the complementation problem of regular languages (see also Sec. 6.2, page 213).

1.3 Thinking in Structures: A Pragmatic Approach to Learning

It is certainly not easy to comprehend, let alone master, Advanced Informatics. Based loosely on what Euclid has been reported to once have said, we can remark that *"There is no silver bullet for problem solving in Advanced Informatics"*. Yet, this does not mean that the journey that leads to the ability to solve problems in Advanced Informatics is unattractive. Rather, this journey offers the achievement of a multitude of milestones and their accompanying feeling of success. This trilogy is expressly designed to nurture early on such a sense of achievement, whilst also introducing throughout further thoughts, methods, and applications from Informatics to provide context for such achievements and to motivate further studies. In this manner, the trilogy weaves several *threads of narrative*. One such thread focuses on stepwise, formal developments oriented towards test-based skills validation; it conveys methods and concepts that are typically used in written exams, but it may well benefit from the additional knowledge and insights provided in the other narrative threads of this trilogy. The remaining threads often project ahead to motivate subsequent material and conceptual developments, and appeal to your intuition and curiosity; these threads are identified by being given a grey background color.

These additional narrative threads are not just meant to motivate, their intent is also to illustrate that the topics of Advanced Informatics are not limited to the aspects pertaining to Theoretical Informatics. Rather, all aspects of Informatics stand to immediately benefit from the patterns of structure-oriented thinking that this trilogy will develop. Such computational and structure-oriented thinking is also an ideal tool for the transfer of methods, knowledge, and technology across the different areas of Informatics itself, and into real-world applications that use Informatics as a means to an end.

Many readers of this trilogy will have as their primary interest the development of skills that are bound to be tested in exams. But those readers should not forget that the material and motivations supplied in the threads given a grey background color will be helpful in several important ways: to illustrate the intent behind methods and solution approaches, to make it possible to relate and use methods and concepts in appropriate ways, and to do so even if the reader may initially struggle to comprehend such methods and concepts.

Our brains often operate in a subconscious way. It is quite common that we understand much later what is meant by something, and that comprehension often arrives in an unexpected context. *"It finally sunk in!"* describes a remarkable phenomenon whose experience can be quite sublime. An illustrating anecdote goes back to Archimedes who allegedly ran through the streets of his hometown Syracuse, completely naked, and repeatedly shouting *"Eureka!"*. This was after he sat in his bathtub where he discovered the sought Principle of Buoyancy (flotation) – in an unexpected but relevant context for studying this problem.

But "it" can only sink in when there is enough material to enable such moments. The numerous references, examples, and pointers provided in this trilogy are meant to stimulate further thinking and discussions, which may well be informed by initial web searches and use of Wikipedia. We believe that most forms of engagement with a topic can aid in its comprehension, perhaps without any immediately visible signs, but without fail.

This examination of Advanced Informatics does not only concern its mathematical topics though. As a matter of fact, it is astonishing what we are able to learn about Informatics through the *Mirror of Mathematics*: this perspective unlocks insights into Informatics that would hardly be attainable through a direct study of Informatics alone. Once this is understood, you have achieved a main aim of this trilogy; then the many parts that it introduces will combine themselves little by little into an organic whole. You may therefore see this trilogy as being designed to be a guide that can accompany your studies so that it may often enable "it" to sink in, and for you to enjoy those moments of sudden realization.

Remarks on using Wikipedia: This trilogy has already made references to entries in Wikipedia and will make more such references subsequently. These references should be understood as provoking more thinking about the discussed concepts. These references are also meant to go deeper into mathematical concepts, to exhibit the complexity of some mathematical problems, and to encourage the acquisition of a general education in Mathematics. Wikipedia entries are therefore a more apt mechanism than referring to appropriate scientific papers and texts. But you should not misunderstand this usage of Wikipedia entries as a supplement for study with this trilogy. **Wikipedia pages are not scientific sources that you can cite as an authority for information in scientific writing**, for example in a seminar article or project report. The reason is that the information you find there is not explicitly vetted by scientists in the respective areas; there is no guarantee that such pages underwent *peer review* for quality assurance. Related to that, there is little control over who edits such pages and in what way. Having said all that, experience shows that Wikipedia pages that are dedicated to topics in Mathematics or Informatics are – by and large – of quite high quality.

This trilogy will expose you to many new concepts, techniques, ways of thinking, and impressions. It is therefore pretty normal that its study will initially make you feel as if you are swimming in a strong current with occasional rapids. It is therefore vital that you then keep swimming so that you will collect enough material for "it" to finally sink in (and so that you won't sink into the water yourself). The moments in time when such Eureka moments happen will differ from reader to reader; some may arrive rather quickly whereas other moments may take longer to appear.

Only if – after serious efforts on your part – you think that you won't achieve the learning outcomes described below, you should rethink your approach to learning and working with this material, perhaps with the support of a mentor or some of your peers. It is also very important to realize that technically demanding texts, such as those found in this trilogy, open up new levels of comprehension and encourage new ways of thinking each time they are read again. The reading of this material in more senior years of study at university, for example, may open connections to material encountered in more junior years or it may provide Eureka moments for topics covered in such junior years. Some topics require appropriate sensibilities and maturity in the discipline, which may not be available at the time they are first covered in a curriculum. This does not only apply to the study of texts. Especially when you encounter topics that were familiar to you from your school days, you should pay close attention to how these topics are now approached. Every level of comprehension allows us to see in a new light that which is already familiar to us. Please be open to moving up to those levels; we hope that you will come to appreciate that there are many such levels worth exploring.

Notational Convention. This trilogy aims to deepen the covered topics and to provide insights into the wider scope of the learned material and its applicability. This is partly achieved through the insertion of numerous textual excursions, forward references, and example discussions. To ensure that you won't lose your narrative flow, especially upon your first reading of a text, we highlighted such additional material with a grey background or within a dedicated section called **Reflections:**

Exploring Covered Topics More Deeply, which you will find at the end of each chapter.

1.4 Learning Outcomes of the First Volume

The leitmotiv of this first volume is to establish the inductive approach as a fundamental principle for system and domain analysis. The study of this volume will make you internalize some elementary mathematical knowledge – for example about sets, propositional logic, relations, and functions – as well as the importance and potential of the explicit separation of syntax and semantics. You, the reader, will become able to construct *compositional models* by combining the consistent application of inductive definitions with the determination of representations (syntax) and meaning (semantics). You will further understand that this principle of compositionality is not only the conceptual foundation of a proof principle that *substitutes equals for equals* – as familiar from school – but that it is at the same time a very pragmatic and perhaps the most effective means of creating structures that *scale* in that they can cope with ever increasing system demands.

An accompanying core principle is that of an *Invariance* and its preservation. You will understand, and gain competencies in, different manifestations of invariances and reasoning about them:

- as invariant patterns within inductive definitions,
- as characteristic traits of induction assumptions,
- as being independent from choice of concrete representation, and
- as structure preservation through functions or transformations (homomorphisms).

The technically most demanding learning outcome for this first volume is that you will gain the competency to use various forms of inductive proofs. You will thus not merely become able to correctly exercise an inductive proof, but you will also acquire the skill of determining which inductive proof method is best for the purpose at hand, and why.

At the intuitive level, you will have taken in that invariances are a key to the mastery and control of change – where change may come in the form of extensions, transformations, or abstractions, and you will have realized that the search for an *Archimedean Point* that is suitable for the scenario under analysis is key to a deeper understanding of that scenario: the more invariances are recognized about a system, the better you will be able to manage its change [73].

Chapter 2
Propositions and Sets

> *A set is a Many that allows itself to be thought of as a One.*
> *(Georg Cantor)*

The aim of Mathematics is to establish abstract, internally consistent constructs of thought. These constructs consist of concepts and their relationships, where the correctness of these relationships is guaranteed by a scientific method of inference. In this method, we start with precise elementary definitions of concepts and relationships (so-called *axioms*). Based on these axioms, we can then infer further properties of these constructs of thought, in order to explore the scope of these constructs and their relationships. The choice of axioms is often motivated by real-world problems. But Mathematics, unlike other natural sciences such as physics, does not necessarily consider constructs that originate from the world in which we live. Axioms may instead model abstract worlds with no immediate connection to our own physical world. Their role and importance may change though over time. A very prominent example is Boolean Algebra, which was hardly recognized by the mathematical community when it was developed. Today, George Boole's work is known as the foundation for much of Computer Science. It will therefore be carefully discussed in the first and second volumes of this trilogy.

Informatics is similar in that regard: it may solve real-world problems such as the creation of a robot that performs basic household functions, or it may invent new ways of distributing and synchronizing systems that have no equivalent in the natural world, but which create a new kind of reality by themselves. The books *Daemon* and *Freedom* by Daniel Suarez [75, 76] provide a good flavor of how much these newly created worlds might affect our lives.

Mathematical constructs of thought seem similar to the *realities* of many computer games, whose *physics* is not necessarily close to our own reality but motivated by visual and other effects that players experience. But even in these virtual worlds, the principle *"What the programmer coded is exactly what happens"*[1] applies here as well, and this principle comes close to the axiomatic approach in Mathematics. This axiomatic approach creates tools in Mathematics and in Informatics that can capture important aspects of reality astonishingly well. Let us mention modern weather

---

[1] In some programming languages, it is possible to write code whose behavior may depend on the used compiler or hardware, or may even be completely undefined. Using this ability in code creates all kinds of problems, for example it may introduce security vulnerabilities.

© Springer International Publishing AG, part of Springer Nature 2018
B. Steffen et al., *Mathematical Foundations of Advanced Informatics*,
https://doi.org/10.1007/978-3-319-68397-3_2

forecasting, the precise control of physical processes such as the flight trajectories of rockets, or the medical imagery of brain scans for diagnostic purposes as examples here. This *status quo* is the result of mathematical traditions that developed and matured over the course of several centuries.

This striving for indisputably correct constructs of thought finds its foremost expression in the quest for provable *propositions* that capture gained insights in a reusable form. The aim of this chapter is to elaborate this concept of propositions and to then put it on firm foundations through the use of elementary set theory. The chapter, and indeed the entire book and trilogy, therefore do not merely convey facts and their understanding. Our ambition is to also practice particular approaches and to nurture a comprehension of the limitations of such approaches, for example through the so-called *antinomies* (see Section 2.3.5) in the case of elementary set theory.

2.1 Propositions

In Mathematics, it is essential that scientific insights be expressed in written form. To reduce the risk of notational ambiguities and the potential misinterpretation of insights or results, mathematicians developed very formal notation and conventions for expressing linguistic constructs.

This need for formulaic notation is perhaps even more pressing in Informatics and Logic, where propositions are often related across several layers of meaning (for example syntax and semantics). Therefore, great care is required in the formalization of propositions as otherwise misunderstandings may come about through the blurring of different layers of meaning. For example, a formula of some logic may occur as a condition in a program, as a side condition and as a loop invariant in a calculus for program verification, as well as (often more informally) at the argumentation level of proofs about programs or calculi themselves. Ensuring the coherent interplay of such levels belongs to the core competencies of an Informatician.

A central concept in all of Logic is that of a *Proposition*, which may be defined as follows.

Definition 2.1 (Propositions). *Propositions* are linguistic constructs expressed in written form to which it is meaningful to assign a truth value *true* (tt) or *false* (ff).
□

When we speak of linguistic constructs above we refer to so-called *declarative sentences*. Something which is not a grammatical sentence, such as *"Cost very not."*, cannot be a proposition. Similarly, imperatives such as "Study the foundations of Informatics!" are not declarative and so cannot be assigned a truth value.

Unfortunately, this definition is not complete in that it may fail to decide whether some declarative sentences are propositions. The problematic aspect of this definition is the term *"meaningful"*. Although it is indubitably meaningful to assign a truth value to many declarative sentences, it is less clear how to do this or whether this is indeed possible for certain declarative sentences. We will make this more precise below and will next list some examples of declarative sentences that certainly are propositions.

Example 2.1 (Propositions).

(1) Dolphins are fish. (*ff*)

(2) The number 5 is a prime. (*tt*)

(3) There are only finitely many prime numbers. (*ff*)

(4) Every even number greater than 2 is the sum of two prime numbers. (?)

(5) Canada is the Olympic Champion in Ice Hockey. (?)

There is not much to say about the first three propositions, except to note that proposition (3) is false, since already Euclid gave a proof of the existence of infinitely many primes.[2] Proposition (4), though, states the so-called *Goldbach's conjecture*[3], whose validity had not been proven at the time of writing this. Nonetheless, it is clear that this declarative sentence is either true or false. We can test, in principle, for each even number $e > 2$ whether e is the sum of two primes: it suffices to test whether $e = p + q$ for some primes p and q less than e.

Proposition (5) is different in the following sense: it is true at the time of writing (August 2015) but it may not be true after the next Olympic Games, and it was false at certain points in the past. Modal and temporal logics were developed in Linguistics and Philosophy [27] and in Informatics [23, 37] to deal with these types of situations.

However, the following are not propositions in the mathematical sense.

Example 2.2 (Not Propositions).

(1) What is the time?

(2) Come here at once!

(3) This proposition is false.

The sentences (1) and (2) are a question and an imperative, respectively, and so are not declarative sentences. In contrast, the sentence (3) is declarative. However, we seem to be unable to consistently assign to it a truth value as it is *self-referential*: if we assign *tt* to it, its statement contradicts this; and we get a similar contradiction if we try to assign the truth value *ff* to it.

Finally, there are declarative sentences for which it is unclear whether or not they should be classified as propositions.

---

[2] See http://en.wikipedia.org/wiki/Euclid's_theorem
[3] See http://en.wikipedia.org/wiki/Goldbach's_conjecture

Example 2.3 (Unclear Propositions).

(1) Today, the weather is fine.

(2) Verdi composed the most significant operas.

On a wet and cold autumn day, most people would find that sentence (1) is false. However, the interpretation of the term *"fine"* is subjective. A tourist in the Scottish Highlands may experience typical seasonal weather there differently from someone who lived there all her life, for example.

Declarative sentence (2) certainly invites us to interpret it through objective means. For example, we may measure the significance of any opera by how often it is performed worldwide in a given calendar year. However, this or any other metric-based approach to assigning a truth value to sentence (2) is likely to raise objections from some musical scholars.

In Informatics and in Mathematics, we consciously avoid such unclear formulations and ambiguities. This is guaranteed by demanding that propositions be presented in a certain precisely defined form (*syntax*), and that there be unique rules for computing unambiguous truth values for such propositions (*semantics*). This method has as an important side effect that the association of a truth value with a – syntactically correct – expression is directly justified as *"meaningful"* via the semantics of that association.

2.1.1 Propositional Logic

In propositional logic, we study how complex propositions can be formed from simpler ones. The idea is that we have certain *elementary* (sometimes called *"atomic"*[4]) propositions which we can compose to create additional, more complex propositions. This approach has two central aspects:

1. *Compositionality:* The structure of elementary propositions is irrelevant for the computation of truth values of complex propositions that contain them: only the truth values of those elementary propositions are needed to complete that computation.
2. *Extensionality Principle:* the truth value of a complex proposition is uniquely determined by the truth values of all of its elementary propositions.

In Chapter 4, especially in Section 4.3, we will make these aspects more formal and precise by inductively defining formulas of propositional logic (*syntax*) and by then assigning to such formulas a meaning (*semantics*) – again using induction. Before we do this, we provide here a less formal approach.

---

[4] Greek `atomos` = "indivisible".

Definition 2.2 (Composition of propositions). Let $\mathscr{A}$ and $\mathscr{B}$ be arbitrary propositions. Then the following constructs are also propositions (also called *formulas of propositional logic*):

- *Negation* of $\mathscr{A}$: written as $\neg\mathscr{A}$, pronounced "not $\mathscr{A}$". The truth value of $\mathscr{A}$ is inverted.[5]

- *Disjunction* of $\mathscr{A}$ and $\mathscr{B}$: written as $\mathscr{A} \vee \mathscr{B}$, pronounced "$\mathscr{A}$ or $\mathscr{B}$". This is true exactly in the cases in which at least one of the two propositions $\mathscr{A}$ and $\mathscr{B}$ is true.

- *Conjunction* of $\mathscr{A}$ and $\mathscr{B}$: written as $\mathscr{A} \wedge \mathscr{B}$, pronounced "$\mathscr{A}$ and $\mathscr{B}$". This is true exactly in the case in which both propositions $\mathscr{A}$ and $\mathscr{B}$ are true.

- *Implication* of $\mathscr{A}$ and $\mathscr{B}$: written as $\mathscr{A} \Rightarrow \mathscr{B}$, pronounced "$\mathscr{A}$ implies $\mathscr{B}$" or "from $\mathscr{A}$ we obtain $\mathscr{B}$". This is true exactly in the cases in which $\mathscr{B}$ is true whenever $\mathscr{A}$ is true.

- *Equivalence* of $\mathscr{A}$ and $\mathscr{B}$: written as $\mathscr{A} \Leftrightarrow \mathscr{B}$, pronounced "$\mathscr{A}$ is equivalent to $\mathscr{B}$" or "$\mathscr{A}$ if, and only if, $\mathscr{B}$". This is true when either both propositions are true or both propositions are false. ☐

In an implication $\mathscr{A} \Rightarrow \mathscr{B}$, we call $\mathscr{A}$ the *premise* and $\mathscr{B}$ the *conclusion* of this implication. You should note that an implication is already true whenever its premise is false. In particular, the proposition:

$$10 \text{ is a prime number } \Rightarrow \text{ elephants can fly}$$

is a true proposition in the world that we live in. From this we learn that we can infer the truth of arbitrary propositions from a proposition that is *logically false*. In Latin, this principle has been known as *ex falso quodlibet*, which we may translate freely as "from falsehood follows whatever one likes."

The operators $\Rightarrow$ (Implication) and $\Leftrightarrow$ (Equivalence) express an "if, then ..." – respectively "if, and only if ..." – relationship between propositions. In normal discourse, the distinctions made in these two relationships are often neglected. For example, the proposition

"If the sun shines today, I will go swimming"

is typically understood to also mean that there won't be any swimming if there is no sunny weather. Although, if we interpret that sentence in formal logic, we cannot really justify such a claim. It is this deviation of everyday linguistic discourse and formal logic that may initially make it harder to treat formal logic with the precision it requires.

The composition symbols $\neg, \vee, \wedge, \Rightarrow$, and $\Leftrightarrow$ are also called *logical operators*. Logical expression that should be evaluated with high priority are enclosed by

---

[5] Occasionally, negation is expressed through an overline bar as in $\overline{\mathscr{A}}$.

parentheses where required. To minimize the use of parentheses in logical expressions, conventions have been established, specifying that some operators have priority over others in their evaluation. These binding priorities say that $\neg$ has highest priority, followed by $\wedge$, followed by $\vee$, followed by $\Rightarrow$ and $\Leftrightarrow$. To see this in an example, the formula $\neg\mathscr{A} \Rightarrow \mathscr{B} \wedge \mathscr{C} \vee \mathscr{D} \Leftrightarrow \mathscr{B}$ really means $(((\neg\mathscr{A}) \Rightarrow ((\mathscr{B}\wedge\mathscr{C})\vee\mathscr{D})) \Leftrightarrow \mathscr{B})$ after we apply this convention.

Remarks on Definition 2.2. Definition 2.2 is remarkable because it is self-referential: although propositions in the formal sense of propositional logic are defined only through this definition, the definition itself already refers to propositions $\mathscr{A}$ and $\mathscr{B}$. For example, for propositions $\mathscr{A}$, $\mathscr{B}$, and $\mathscr{C}$ the expressions $\neg\mathscr{A}$ and $(\mathscr{B}\vee\mathscr{C})$ are also propositions. And therefore, $\neg(\mathscr{B}\vee\mathscr{C})$ is a proposition as well. This is an example of an *inductive definition*,[a] a concept we will formalize in Chapter 4. Two properties of such inductive definitions are immediate: firstly, the set of propositional logic formulas is unbounded, because we can always form more complex propositions from propositions we have already constructed. Secondly, it is easy to see why we need elementary propositions. Otherwise, the definition would have no initial points on which to base its constructions for creating propositions.

We call structures that can be built in finitely many steps from atomic building blocks *well founded*. Almost all of the structures that we will consider in this book are well founded and, therefore, can be reasoned about with inductive proofs. Further details of such structures and their proof principles are found in Chapter 5.

The meaning of the logical operators is defined by formulations that also capture such logical relationships. For example, the formulation "*if, and only if . . .*" expresses an equivalence relationship. Because of its importance, it is often abbreviated by *iff*. This illustrates the importance of the separation of different levels of meaning, which we already highlighted in the introduction. The equivalence "$\Leftrightarrow$" is a syntactic operator in propositional logic (and so provides a case in the inductive definition of these formulas). Whereas the statement "*if, and only if . . .*" takes place on a level at which we compare the meaning of two formulas of propositional logic. In linguistics, we could say that "$\Leftrightarrow$" is an expression of the object language (formulas of propositional logic), whereas "*if, and only if . . .*" is an expression of the meta-language (which considers relationships between expressions of the object language, in this case the intended meaning of the logical operators of the object language). In particular, we cannot use these expressions interchangeably!

---

[a] Alternatively, one may also speak of a *recursive* definition here.

Truth tables provide a formal method for evaluating propositions according to the meaning of propositional logic given in Definition 2.2.

$\mathscr{A}$	$\mathscr{B}$	$\neg\mathscr{A}$	$\mathscr{A} \vee \mathscr{B}$	$\mathscr{A} \wedge \mathscr{B}$	$\mathscr{A} \Rightarrow B$	$\mathscr{A} \Leftrightarrow \mathscr{B}$
ff	*ff*	*tt*	*ff*	*ff*	*tt*	*tt*
ff	*tt*	*tt*	*tt*	*ff*	*tt*	*ff*
tt	*ff*	*ff*	*tt*	*ff*	*ff*	*ff*
tt	*tt*	*ff*	*tt*	*tt*	*tt*	*tt*

Table 2.1 Truth tables for all logical operators of propositional logic

The columns of a truth table are separated into two zones: on the left are columns for the elementary propositions of the formulas whose truth tables we wish to compute (here $\mathscr{A}$ and $\mathscr{B}$); on the right, separated by $\|$, are truth tables for composed formulas. Note that a row in such tables specifies a unique *assignment* of truth values to all elementary propositions, and corresponding truth values for all formulas under consideration. This notion is important enough to deserve a definition.

Definition 2.3 (Assignment). Let $\mathscr{A}$ be a formula of propositional logic that contains $k \geq 1$ atomic propositions. An *assignment* for formula $\mathscr{A}$ associates with all atomic propositions occurring in $\mathscr{A}$ a truth value *tt* or *ff*. □

From this definition we learn that a proposition with k atomic propositions has 2^k different assignments. We will reflect on truth tables and their use further in Section 2.3.1.

DNF and functional completeness. Looking at Table 2.1, it is clear that we can compute complete columns for all formulas whose elementary propositions have matching columns on the left. Now let us do a *Gedankenexperiment*: suppose that we create a column on the right and fill it with arbitrary truth values. Is it then possible to *synthesize* a formula over the elementary propositions whose column would equal the one we just created without referring to any formula?

It turns out that this is indeed possible. One way of seeing this is through a construction that relies on so-called *disjunctive normal forms* (DNF). For example, let us assume that we no longer know that the column for formula $\mathscr{A} \Leftrightarrow \mathscr{B}$ in Table 2.1 is really for formula $\mathscr{A} \Leftrightarrow \mathscr{B}$. We can construct an equivalent formula from that column as follows: for each entry *tt* in that column, look at the truth values of elementary propositions in that row and form a conjunction that captures this set of truth values; then form the disjunction of all such conjunctions. In our example, the first and fourth row have truth value *tt* and give rise to disjunctions $\neg\mathscr{A} \wedge \neg\mathscr{B}$ and $\mathscr{A} \wedge \mathscr{B}$, respectively. The DNF for this column is therefore $(\neg\mathscr{A} \wedge \neg\mathscr{B}) \vee (\mathscr{A} \wedge \mathscr{B})$.

It is intuitively clear that these formulas only involve the logical operators $\neg$, $\wedge$, and $\vee$. We can therefore say that the set of operators $\{\neg, \wedge, \vee\}$ is *functionally complete*.[a] Furthermore, it is clear that the shape of these formulas is such that they are a disjunction of conjunctions, and where the conjunctions only have elementary propositions or their negations as arguments. DNFs are formulas that have exactly this shape.

---
[a] This fact will be formalized and proven in Section 5.5.

Propositions $\mathscr{A}$ and $\mathscr{B}$ that have identical columns in their truth tables are called *(semantically) equivalent*, and we formally express this equivalence as $\mathscr{A} \equiv \mathscr{B}$. For example, we can derive semantic equivalences that show how the logical operators for conjunction, implication, and logical equivalence can be expressed with the operators $\neg$ and $\vee$ alone.

$$\mathscr{A} \wedge \mathscr{B} \equiv \neg(\neg\mathscr{A} \vee \neg\mathscr{B}) \tag{2.1}$$

$$\mathscr{A} \Rightarrow \mathscr{B} \equiv \neg\mathscr{A} \vee \mathscr{B} \tag{2.2}$$

$$\mathscr{A} \Leftrightarrow \mathscr{B} \equiv \neg(\neg(\neg\mathscr{A} \vee \mathscr{B}) \vee \neg(\neg\mathscr{B} \vee \mathscr{A})) \tag{2.3}$$

The semantic equivalences (2.1) – (2.3) show that negation and disjunction alone suffice to express all logical operators introduced in Definition 2.2. This is hardly surprising: we already saw that the set of logical operators $\{\neg, \wedge, \vee\}$ is functionally complete. And since $\wedge$ is expressible through $\neg$ and $\vee$ as in the semantic equivalence (2.1), it follows naturally that the set of logical operators $\{\neg, \vee\}$ is functionally complete as well. In particular, we can express implication and logical equivalence through $\neg$ and $\vee$ alone – as seen in (2.2) and (2.3)[6].

NAND and XOR The logical operator "NAND", denoted by $\overline{\wedge}$, can be defined through $\mathscr{A} \overline{\wedge} \mathscr{B} =_{df} \neg(\mathscr{A} \wedge \mathscr{B})$. This operator thus captures the negation of the conjunction of its two arguments. It turns out that $\{\overline{\wedge}\}$ is already functionally complete: since $\{\neg, \vee\}$ is functionally complete, this claim follows as $\overline{\wedge}$ can express both $\neg$ and $\vee$: we have the semantic equivalences $\neg\mathscr{A} \equiv \mathscr{A} \overline{\wedge} \mathscr{A}$ and $\mathscr{A} \vee \mathscr{B} \equiv \neg\mathscr{A} \overline{\wedge} \neg\mathscr{B} \equiv (\mathscr{A} \overline{\wedge} \mathscr{A}) \overline{\wedge} (\mathscr{B} \overline{\wedge} \mathscr{B})$.

The functional completeness of $\{\overline{\wedge}\}$ was particularly important in circuit design, as it allowed one to construct arbitrary circuits from uniform and mass-produced logic gates. Today's chip production techniques allow for the *burning* of entire complex systems onto a piece of hardware, a concept known as *System on a Chip (SoC)*. Therefore, the significance of

$$universal \tag{2.4}$$

logic gates is diminishing although they still play a role, for example in Flash storage modules.

Another logical operator is the *XOR*, pronounced *exclusive OR* and formally expressed with the symbol $\oplus$. This operator is less significant in Mathematical Logic but plays an important role in Informatics, in particular in

---
[6] The formal proof of functional completeness of $\{\neg, \vee\}$ is the subject of Section 5.5.

> Cryptography. The proposition $\mathscr{A} \oplus \mathscr{B}$ is true when exactly one of $\mathscr{A}$ and $\mathscr{B}$ is true. It is an instructive exercise to show the semantic equivalence $\mathscr{A} \oplus \mathscr{B} \equiv \neg(\mathscr{A} \Leftrightarrow \mathscr{B})$.
>
> The operator $\oplus$ alone is not functionally complete. But it has a remarkable algebraic property that can be expressed in the semantic equivalence $(\mathscr{A} \oplus \mathscr{B}) \oplus \mathscr{B} \equiv \mathscr{A}$. In other words, if we apply $\oplus$ twice with the same argument, then these two applications cancel each other out. This is important in Cryptography if we think of each application of $\oplus \mathscr{B}$ as modeling encryption and decryption of data.

Let us stress that semantic equivalence $\equiv$ is *not* a logical operator of propositional logic. We can compare it to "if, and only if ...", which is a meta-symbol that expresses a proposition about propositions. However, the meta-symbol $\equiv$ and the logical operator $\Leftrightarrow$ are consistent with each other: whenever we have the semantic equivalence $\mathscr{A} \equiv \mathscr{B}$, it then follows that the formula $\mathscr{A} \Leftrightarrow \mathscr{B}$ is always true (its truth table column only has entries tt). Such formulas are called *tautologies*. To express this in other words, $\mathscr{A} \equiv \mathscr{B}$ is a proposition at the level of argumentation (meta-language), whereas $\mathscr{A} \Leftrightarrow \mathscr{B}$ is a logically identical proposition at the object level.

The semantic equivalences (2.1) and (2.2) on page 26 are shown in the following truth tables.

$\mathscr{A}$	$\mathscr{B}$	$\neg\mathscr{A}$	$\neg\mathscr{B}$	$\neg\mathscr{A} \vee \neg\mathscr{B}$	$\neg(\neg\mathscr{A} \vee \neg\mathscr{B})$	$\mathscr{A} \wedge \mathscr{B}$	$\neg\mathscr{A} \vee \mathscr{B}$	$\mathscr{A} \Rightarrow \mathscr{B}$
$f\!f$	$f\!f$	tt	tt	tt	$f\!f$	$f\!f$	tt	tt
$f\!f$	tt	tt	$f\!f$	tt	$f\!f$	$f\!f$	tt	tt
tt	$f\!f$	$f\!f$	tt	tt	$f\!f$	$f\!f$	$f\!f$	$f\!f$
tt	tt	$f\!f$	$f\!f$	$f\!f$	tt	tt	tt	tt

Table 2.2 Truth tables that show some semantic equivalences

This example suggests an important proof principle for propositional logic, which captures the ability to compute truth tables *incrementally*:

Proof Principle 1 (Incremental Truth Tables)

Let $\mathscr{A}$ and $\mathscr{B}$ be formulas of propositional logic. Given truth tables for formulas $\mathscr{A}$ and $\mathscr{B}$, we may construct truth tables for $\neg\mathscr{A}$, $\mathscr{A} \wedge \mathscr{B}$, $\mathscr{A} \vee \mathscr{B}$, $\mathscr{A} \Rightarrow \mathscr{B}$, and $\mathscr{A} \Leftrightarrow \mathscr{B}$ by only referring to the truth tables of $\mathscr{A}$ and $\mathscr{B}$, not to truth tables of atomic propositions or any other sub-formulas of $\mathscr{A}$ and $\mathscr{B}$.

In this text, we will highlight practically important proof principles in the manner in which we did this for incremental truth tables above. This proof principle is also applied in the truth tables of Table 2.2. The truth table for formula $\neg(\neg\mathscr{A} \vee \neg\mathscr{B})$, for example, is not computed from scratch, but only inspects the truth table for $\neg\mathscr{A} \vee \neg\mathscr{B}$ and negates its values.

In Lemma 2.1 below, we list semantic equivalences that are very useful when it comes to simplifying formulas of propositional logic without changing their meaning. Some of these equivalences make use of the symbols T and F, which express elementary propositions that are always true, respectively always false. We leave the proof of the semantic equivalences listed in that lemma as an exercise.

Lemma 2.1 (Semantic Equivalences). *Let $\mathscr{A}, \mathscr{B}, \mathscr{C}$ be arbitrary propositions. Then the following semantic equivalences hold:*

$$\mathscr{A} \wedge \mathscr{B} \equiv \mathscr{B} \wedge \mathscr{A} \qquad\qquad (Commutativity)$$
$$\mathscr{A} \vee \mathscr{B} \equiv \mathscr{B} \vee \mathscr{A}$$

$$(\mathscr{A} \wedge \mathscr{B}) \wedge \mathscr{C} \equiv \mathscr{A} \wedge (\mathscr{B} \wedge \mathscr{C}) \qquad\qquad (Associativity)$$
$$(\mathscr{A} \vee \mathscr{B}) \vee \mathscr{C} \equiv \mathscr{A} \vee (\mathscr{B} \vee \mathscr{C})$$

$$\mathscr{A} \wedge (\mathscr{A} \vee \mathscr{B}) \equiv \mathscr{A} \qquad\qquad (Absorption)$$
$$\mathscr{A} \vee (\mathscr{A} \wedge \mathscr{B}) \equiv \mathscr{A}$$

$$\mathscr{A} \wedge (\mathscr{B} \vee \mathscr{C}) \equiv (\mathscr{A} \wedge \mathscr{B}) \vee (\mathscr{A} \wedge \mathscr{C}) \qquad\qquad (Distributivity)$$
$$\mathscr{A} \vee (\mathscr{B} \wedge \mathscr{C}) \equiv (\mathscr{A} \vee \mathscr{B}) \wedge (\mathscr{A} \vee \mathscr{C})$$

$$\mathscr{A} \wedge \neg\mathscr{A} \equiv \mathsf{F} \qquad\qquad (Negation)$$
$$\mathscr{A} \vee \neg\mathscr{A} \equiv \mathsf{T}$$

$$\mathscr{A} \wedge \mathscr{A} \equiv \mathscr{A} \qquad\qquad (Idempotency)$$
$$\mathscr{A} \vee \mathscr{A} \equiv \mathscr{A}$$

$$\neg\neg\mathscr{A} \equiv \mathscr{A} \qquad\qquad (Double\ Negation)$$

$$\neg(\mathscr{A} \wedge \mathscr{B}) \equiv \neg\mathscr{A} \vee \neg\mathscr{B} \qquad\qquad (De\ Morgan's\ Laws)$$
$$\neg(\mathscr{A} \vee \mathscr{B}) \equiv \neg\mathscr{A} \wedge \neg\mathscr{B}$$

$$\mathsf{T} \wedge \mathscr{A} \equiv \mathscr{A} \qquad\qquad (Neutrality)$$
$$\mathsf{F} \vee \mathscr{A} \equiv \mathscr{A}$$

We may take the semantic equivalences of Lemma 2.1 as the basis for an axiomatic proof system for propositional logic, a topic we will explore in Section 2.1.3 in more detail.

Negation normal form. The so-called *De Morgan's Laws* can be put to use to push the negation of a binary logical operator into the arguments of that operand. Since the only non-binary operator is logical negation, its negation can be eliminated through *Double Negation*. Therefore, we may use these semantic equivalences to convert a formula of propositional logic into a semantically equivalent one in which negations are only applied to atomic propositions. This process is sometimes referred to as *negation pushing*, and its output is called *negation normal form* (NNF) .

For example, consider the formula $\neg(\neg\mathscr{A} \wedge (\mathscr{B} \vee \mathscr{C}))$ of propositional logic. If we push the outermost negation inwards, this turns into $\neg\neg\mathscr{A} \vee \neg(\mathscr{B} \vee \mathscr{C})$, and double negation elimination transforms this into $\mathscr{A} \vee \neg(\mathscr{B} \vee \mathscr{C})$. Pushing the last negation inwards then results in the NNF $\mathscr{A} \vee (\neg\mathscr{B} \wedge \neg\mathscr{C})$, which by construction is semantically equivalent to $\neg(\neg\mathscr{A} \wedge (\mathscr{B} \vee \mathscr{C}))$. To apply negation pushing to the operators $\Rightarrow$ and $\Leftrightarrow$, we first have to apply semantic equivalences that replace these operators; for example the equivalence $\mathscr{A} \Rightarrow \mathscr{B} \equiv \neg\mathscr{A} \vee \mathscr{B}$ that gets rid of the implication operator.

Note that the functional completeness of $\{\neg, \vee\}$ (see page 26) does not directly imply an algorithm for negation pushing. Nor does it provide a proof that negation pushing always produces a negation normal form. It turns out that such a proof is technically more challenging than showing the functional completeness of $\{\neg, \vee\}$ and this is pursued further in Section 5.7.1.

The functional completeness argument came with an explicit recipe for how to convert any formula of propositional logic into a semantically equivalent one that uses only operators $\neg$ and $\vee$. It is intuitively clear that this recipe can be turned into an algorithm that automatically computes these conversions. In Mathematics and in Informatics such explicit proofs are called *constructive*, and we will explore this concept further on page 41.

Historically, the most significant application of propositional logic is in the design and analysis of digital circuits. But modern applications of propositional logic are found in many different areas and subjects, for example in reliability and safety analysis of critical systems. We refer to Section 2.3.1 for further reflections on such applications.

2.1.2 Predicate Logic

Propositional logic may well suffice to model digital circuits, help with identifying possible diseases given a set of symptoms, and so forth. But there are many practical problems in which the internal structure of atomic propositions is of great importance and so needs to be modeled with more accuracy. For example, the atomic propositions

$$1+2 = 2+1$$
$$2+3 = 3+2$$
$$6+14 = 14+6$$
$$\vdots$$

all follow the same *pattern*. Namely, they are instances of the commutativity of the addition of two numbers. For the integers already, there will be infinitely many such patterns and so we cannot possible state them all. To capture such patterns completely, predicate logic therefore introduces *logical quantifiers* such as "*for all*" and "*there exists*" into the syntax of its logic.

Let us write $\mathbb{N}$ for the set of natural numbers.[7] With that notation, we can now express the commutativity of addition over the natural numbers in the following formula of predicate logic:

$$\forall x \in \mathbb{N}. \; \forall y \in \mathbb{N}. \; x+y = y+x$$

You may think of this formula as declaring two variables x and y of type $\mathbb{N}$, and then stating that all instances of these variables over their types satisfy the equation $x+y = y+x$. We hasten to note that a formal foundation of predicate logic will not be required for appreciating the material of this trilogy. Therefore, we refer the interested reader to other texts such as [37, 32] for more about such foundations.

An integral part of predicate logic is an underlying structure over which a formula of that logic may be interpreted. Such a structure consists of a non-empty set of *individuals*, with corresponding *operations* and *relations* (see Section 3.1).

An important example of such a structure is the set $\mathbb{N}$ of natural numbers as set of individuals, the operations of addition and multiplication defined on that set, and the relations $=$ (equality) and $\leq$ (less-than-or-equal) that can relate two natural numbers with the obvious interpretation. Predicate logic over the natural numbers extends propositional logic in the following manner:

1. The logical operators of propositional logic may still be used in the manner familiar from propositional logic.

2. Expressions may be used that contain within them free variables over the natural numbers, for example $n+1 \leq 3$ with free variable n. Such expressions are referred to as *predicates*, and we write $\mathscr{A}(x_1,\ldots,x_n)$ to indicate that predicate $\mathscr{A}$ contains $x_1, \ldots, x_n$ all as free variables.[8] Predicates may contain several variables, for example $n = p \cdot q$.

3. Using *logical quantifiers*, we can transform predicates into formulas of predicate logic. Concretely, for a proposition $\mathscr{A}(n)$ with free variable n we may form

---

[7] In this trilogy, we assume that zero 0 is a natural number, and we write $\mathbb{N}_{>0}$ to denote the set of positive natural numbers, which does not contain 0.

[8] We can determine a truth value of a predicate when its free variables are associated with individuals. For example, predicate $n+1 \leq 3$ is true for association $n \mapsto 2$, but false for association $n \mapsto 3$.

- *Universal Quantification* over $\mathscr{A}$: the formula $\forall n. \mathscr{A}(n)$
  This universal quantification is true if, and only if, proposition $\mathscr{A}(n)$ is true for all values n in $\mathbb{N}$. In fact, $\forall n. \mathscr{A}(n)$ can be regarded as a finite representation of a specific infinite conjunction.

- *Existential Quantification* over $\mathscr{A}$: the formula $\exists n. \mathscr{A}(n)$
  This existential quantification is true, if, and only if, proposition $\mathscr{A}(n)$ is true for at least one value n in $\mathbb{N}$. In fact, $\exists n. \mathscr{A}(n)$ can be regarded as a finite representation of a specific infinite disjunction.

The meaning of quantification for structures other than $\mathbb{N}$ is defined similarly; quantification is then over the elements of *that particular* structure. If the context won't determine that the structure is that of the natural numbers, we may also write the formulas $\forall n \in \mathbb{N}. \mathscr{A}(n)$, respectively $\exists n \in \mathbb{N}. \mathscr{A}(n)$, to highlight the structure used in the quantification. In a quantified formula, variable n is *bound* to its quantifier, and so cannot be referenced by outer quantifiers.[9] For example, the formula $\exists n. \forall n. n = n+0$ is always true over our structure of natural numbers, even though the existential quantifier cannot reference any of the n variables in predicate $n = n+0$.

Quantifiers do not necessarily have to be followed by a predicate. For example, let $\mathscr{B}$ be a predicate with two free variables. Then $\forall n. \exists m. \mathscr{B}(n,m)$ as well as the expression $\exists m. (\mathscr{B}(m,m) \vee \mathscr{B}(2 \cdot m,m))$ are formulas of predicate logic.

Of course, one may define predicates over structures other than that of the natural numbers. For example, consider the formula $\exists x \in \mathbb{R}. \forall y \in \mathbb{N}. \neg(x = y)$ which expresses that there is a real number that is not equal to any natural number. In particular, predicates and formulas may mix elements from different structures.

As a convention of binding priorities, quantifiers have lower binding priority than logical operators. This convention aids human readability but may have to be overwritten by the insertion of parentheses to capture the intended effect in a formula. For example, formula $\forall x. (x < 5) \vee (x = 1)$ does not contain any free variables, whereas the formula $(\forall x. (x < 5)) \vee (x = 1)$ contains the rightmost occurrence of x as a free variable.

Instead of $\neg(\forall n. \mathscr{A}(n))$, we may write $\not\forall n. \mathscr{A}(n)$; and we may similarly abbreviate $\neg(\exists n. \mathscr{A}(n))$ by $\not\exists n. \mathscr{A}(n)$. Nested quantification with the same type (universal/existential) of quantifier, for example the nested universal quantification $\forall x_1. \forall x_2. \ldots \forall x_n. \mathscr{A}(x_1, \ldots, x_n)$, may be abbreviated by $\forall x_1, \ldots, x_n. \mathscr{A}(x_1, \ldots, x_n)$.

We can appreciate the expressive power of predicate logic by noting that relations may be defined through complex formulas, and that these relations in turn may then be used in formulas. This approach does not really increase the expressiveness of the logic, but it encourages a modular use and reuse of formulas, and it makes complex formulas more succinct and easier to read by humans. We illustrate this by means of the properties of the *greatest common divisor* (gcd).

---

[9] This form of binding follows the same principle as that of *static scoping* of locally declared variables as used in many programming languages.

Such modular definitions of relations often use auxiliary relations and their definitions, much in the way in which methods or functions in programming languages are often defined in terms of other methods or functions. Here we use the auxiliary *divisibility relation* $|$, which takes two arguments and which we write in so-called *infix* notation – meaning that the relation is written in the middle of the two arguments:

$$n|m \ =_{df} \ \exists k \in \mathbb{N}. \ n \cdot k = m$$

In words this says that n divides m if there is a natural number k such that the k-fold product of n equals the number m. For example, $2|12$ holds since we may choose $k = 6$, but $3|16$ does not hold since $3 \cdot k \neq 16$ for *all* k in $\mathbb{N}$.

From this binary relation, we can define in predicate logic a ternary relation (meaning that this relation has three arguments) which specifies when a number x is the greatest common divisor of numbers n and m:

$$gcd(n,m,x) \ =_{df} \ x|n \ \wedge \ x|m \ \wedge \ \forall y \in \mathbb{N}. \ (y|n \ \wedge \ y|m) \ \Rightarrow \ y \leq x \qquad (2.5)$$

When using infix notation, we assume that relations have higher binding priority than logical operators. Since $\wedge$ has higher priority than $\Rightarrow$, and since $\Rightarrow$ has higher priority than logical quantification, the formula that defines relation $gcd(n,m,x)$ is a conjunction. The first two conjuncts, $x|n \ \wedge \ x|m$, demand that x be a common divisor of n and m. The third conjunct is a universally quantified formula which states that x is greater than or equal to all common divisors of n and m. Together, these three conjuncts therefore correctly capture that x is the greatest common divisor of n and m.

It is possible to view the ternary relation gcd as a function (see Section 3.2 for a precise definition of functions) of type $\mathbb{N} \times \mathbb{N} \to \mathbb{N}$: it takes two natural numbers n and m as input, and outputs a unique natural number x. This justifies writing $x = gcd(n,m)$ instead of $gcd(n,m,x)$ if so desired.

However, this mathematical definition of the function gcd would not be satisfactory for specifying a function in Informatics. For the latter, we also need a description of the computation that the function has to perform, not just its type and expected input/output behavior. This illustrates a fundamental difference between Mathematics and Informatics. Note that the above definition of gcd contains no information about how to compute it. Similarly, the definition of the auxiliary divisibility relation contains no explicit description of how to determine whether an instance $n|m$ is true and, if so, how a k may be found that witnesses this fact.

Suppose we have a claimed implementation of function gcd. How can we formally *verify* that this implementation meets the specification in (2.5) of function gcd? The need for formal verification (an activity at a meta-level) means that the specification (given at the object level) also has to be expressed formally: we cannot hope to formally prove relationships based on informal descriptions of said relationships. Such formal specifications, for example expressed in predicate logic, are then a basis for determining a suitable Archimedean Point, i.e., an *invariant* which is sufficient for establishing the desired correctness of the implementation [73].

Let us now consider invariant-based reasoning by means of a concrete example, *Euclid's Algorithm* for the computation of the greatest common divisor – which we introduced in Chapter 1 already.

The way in which this algorithm operates is really simple: it keeps substracting the smaller number from the larger one until the larger one becomes the smaller one (the modulo operation); and then it repeats this process until one of the numbers becomes 0. When this happens, the other number equals the *gcd* of the initial input to that algorithm. That this really computes the *gcd* relies on the following interconnected reasons:

- *Invariance:* The *gcd* of a and b equals the *gcd* of $a - b$ and b if $a > b$. Therefore, we may compute the *gcd* of a and b by computing the *gcd* of $a - b$ and b; and we may continue this simplification process as long as we can be sure that $a > b$ remains true (for example by swapping these arguments if need be). This invariant is captured in Theorem 2.1 below.
- *Termination:* Euclid's Algorithm stops after a finite number of iterations. To see this, note that the sum of its arguments, $a + b$, strictly decreases with each iteration but stays non-negative. This obviously can only happen finitely often. Specifically, either b is already 0 and so the algorithm stops there; or $a > b > 0$ and so $a - b$ is less than a, meaning that the sum $b + (a - b)$ of the next iteration is less than the sum $a + b$ of the previous one. A formal treatment of termination can be found in Section 5.3.1.
- *Consequence:* The algorithm has the stated invariant as property and it terminates such that one of the two numbers is 0. From this we infer that the algorithm is correct: it outputs two numbers, n_T and 0 (by termination), and the *gcd* of n_T and 0 is the *gcd* of the original input values for a and b (by invariance). Since $0 = 0 \cdot n$, we see that every natural number n is a divisor of 0. In particular, n_T is a divisor of 0. Since n_T is also a divisor of itself, it follows that n_T is the *gcd* of n_T and 0, and therefore, by invariance, of the original values a and b as well.

Theorem 2.1. *Let m, n be natural numbers with $m < n$. Then the following hold:*

1. *For every divisor x of m, we have $x|n$ if, and only if, $x|(n - m)$.*
2. *The gcd of m and n equals the gcd of $n - m$ and m.*

Proof. 1. Since $x|m$, there is a natural number k with $x \cdot k = m$. We prove the implications addressed by the phrase *if, and only if* separately.

"$\Rightarrow$":   Since $x|n$, there is a natural number k' with $x \cdot k' = n$. Therefore we have

$$n - m = (x \cdot k') - (x \cdot k) = x \cdot (k' - k)$$

Since $m < n$, we infer $k \leq k'$ and so $(k' - k)$ is in $\mathbb{N}$, from which $x|(n - m)$ follows.

"$\Leftarrow$":   Since $x|(n-m)$, there is a natural number k' with $x \cdot k' = (n-m)$. Therefore we have

$$n = (n-m) + m = (x \cdot k') + (x \cdot k) = x \cdot (k' + k)$$

from which $x|n$ follows since $k' + k$ is clearly in $\mathbb{N}$.

2. Let $x = gcd(m,n)$ and $x' = gcd(n-m,m)$. We need to show $x = x'$, which we can show by proving both $x \leq x'$ and $x' \leq x$:

   a. To show $x \leq x'$ it suffices to show that x is a common divisor of $n-m$ and m, since x' is the greatest such common divisor. Since x is by definition a common divisor of m and n, we know by Theorem 2.1(1) that x is also a divisor of $n-m$.

   b. To show that $x' \leq x$, it suffices to show that x' is a common divisor of m and n, since x is the greatest such divisor. Since x' is by definition a common divisor of $n-m$ and m, there are natural numbers k_1 and k_2 such that $n-m = k_1 \cdot x'$ and $m = k_2 \cdot x'$. From this we infer that $n = (n-m) + n = k_1 \cdot x' + k_2 \cdot x' = (k_1 + k_2) \cdot x'$, which shows that x' is a divisor of n as $k_1 + k_2$ is a natural number.

$\square$

Compositionality We used a parameterized predicate such as $gcd(a,b,x)$ to express a complex condition in terms of simpler ones such as $x|a$. This illustrates the effect of *compositionality* on the construction of complex specifications from simple building blocks. This example also illustrates that compositionality allows us to write – with confidence – derived, complex specifications that we could hardly write with high confidence if we had to fully expand all their atomic propositions.

"*Derived*" here means *expressed through constructs of the base language* (here predicate logic). It is in this sense that procedures, methods, or functions of programming languages are derived from the functionalities of assignments and basic control structures: we could replace an invocation of a function, for example, with the appropriately instantiated copy of its definitional body, much like we could expand expression $gcd(n,m,x)$ via the right-hand side of the equational definition in (2.5). This is in contrast to the derivation of the concept of a procedure itself, which is a genuine language extension. An example of this would be to add procedures to the WHILE-language [56]. We refer to [67] for a systematic study of extension principles in the design of programming languages.

The introduction of derived, parameterized predicates, however, allows us to extend the base logic (in this case predicate logic) to successively more expressive (domain-specific) specification languages. This leads to a library-based approach, which is also familiar from most modern programming languages. Indeed, the language Scala [59] is expressly designed to support the successive extension of its language to make it more domain-specific.

A frequent source of misunderstandings is the effect that logical negation has on logical quantification; this effect is not necessarily consistent with the use of negation and quantification in everyday linguistic discourse. For example, the negation of "*all*" is not "*none*", as one might expect, but rather "*not all*".

Formally, we have the following relationships.

Lemma 2.2 (Negation of quantified formulas). *Let $\mathscr{A}$ be a proposition with one free variable x. Then the following semantic equivalences hold:*

1. $\neg(\forall x.\ \mathscr{A}(x)) \equiv \exists x.\ \neg\mathscr{A}(x)$

2. $\neg(\exists x.\ \mathscr{A}(x)) \equiv \forall x.\ \neg\mathscr{A}(x)$

Proof. For sake of illustration, we show the first equivalence above, where we assume that the structure is already given and understood; for example, it could be the natural numbers. It is important to note that the proof below works for all choices of such structures:

	$\neg(\forall x.\ \mathscr{A}(x))$	is true	
iff	$\forall x.\ \mathscr{A}(x)$	is false	(Evaluation of $\neg$)
iff	$\mathscr{A}(x)$	is false for at least one x	(Evaluation of $\forall$)
iff	$\neg\mathscr{A}(x)$	is true for at least one x	(Evaluation of $\neg$)
iff	$\exists x.\ \neg\mathscr{A}(x)$	is true	(Evaluation of $\exists$) □

Negation normal form for predicate logic. Recall *negation pushing* for formulas of propositional logic from page 28. We can extend this method to formulas of predicate logic as follows: when pushing negations into the formula we pretend that all predicates such as $x < y$ are atomic propositions, and so negation won't be pushed into these predicate expressions; and we use the semantic equivalences of Lemma 2.2 to push negations into logical quantification.

For example, consider the formula $\neg((\forall x.\ \exists y.\ x \leq y) \wedge (\neg \exists z.\ z < 0))$. Pushing the outermost negation inwards we obtain $\neg(\forall x.\ \exists y.\ x \leq y) \vee (\neg\neg \exists z.\ z < 0)$. The double negation gets eliminated so that we then obtain $\neg(\forall x.\ \exists y.\ x \leq y) \vee (\exists z.\ z < 0)$. The remaining negation first passes through the universal quantification to render $(\exists x.\ \neg \exists y.\ x \leq y) \vee (\exists z.\ z < 0)$, and then negation passes through the existential quantification for y to yield $(\exists x.\ \forall y.\ \neg(x \leq y)) \vee (\exists z.\ z < 0)$. This is a formula in negation normal form. Note that negation pushing won't transform $\neg(x \leq y)$ into $y < x$; and doing so would change the meaning of the formula for some interpretations of $\leq$ and $<$ in *partial orders*, a subject studied in the second volume of this trilogy.

Negation pushing is an important technique in the area of fixed-point logics [23, 51], as it allows us to control if not eliminate the only non-monotonic logical operator.[a] Such control leads to more precise or more efficient reasoning techniques.[b] Just as invariants in programs can be seen as Archimedean points, we may think of monotonicity, i.e., the preservation of order, as the Archimedean point in fixed-point logics: monotonicity implies that the more we know, the more we may infer from that knowledge.

---

[a] Monotonic logical operators preserve implication: if one replaces the operands of a monotonic operator by operands that are implied by the original operators, then the new composite expression is also implied by the original expression.

[b] We will study monotonicity in the second volume, *Algebraic Thinking*.

2.1.3 Approaches and Principles of Logical Proof

We already saw that truth tables are a sufficient means for proving arbitrary theorems in propositional logic. But this method has practical limits: a formula with k atomic propositions in it has a truth table with 2^k rows. For predicate logic, this limitation of truth tables becomes even more pronounced since structures may have infinitely many individuals, for example the natural numbers as elements of $\mathbb{N}$. We cannot compute tables with infinitely many rows. Therefore, we seek alternative means of reasoning about formulas of predicate logic and two approaches for such reasoning have established themselves.

For one, we may construct a proof *semantically*, and this approach appeals directly to the definitional meaning of logical operators and quantifiers. An example of this approach can be seen in the proof of Lemma 2.2.[10]

For another, we may apply rules that syntactically manipulate formulas step by step to prove a desired theorem. These rules would already have been shown independently, for example through semantic proofs. The application of rules to transform formulas renders *syntactic* proofs and this approach is most fruitfully pursued in automated-reasoning tools, for example in Satisfiability Modulo Theories solvers [7]. To see an example of a syntactic proof, we show the semantic equivalence $\top \lor \mathscr{A} \equiv \top$ by using only rules that were stated to be valid in Lemma 2.1:

$$
\begin{aligned}
\top \lor \mathscr{A} &\equiv (\mathscr{A} \lor \neg\mathscr{A}) \lor \mathscr{A} && \text{(Negation)} \\
&\equiv \mathscr{A} \lor (\neg\mathscr{A} \lor \mathscr{A}) && \text{(Associativity)} \\
&\equiv \mathscr{A} \lor (\mathscr{A} \lor \neg\mathscr{A}) && \text{(Commutativity)} \\
&\equiv (\mathscr{A} \lor \mathscr{A}) \lor \neg\mathscr{A} && \text{(Associativity)} \\
&\equiv \mathscr{A} \lor \neg\mathscr{A} && \text{(Idempotency)} \\
&\equiv \top && \text{(Negation)}
\end{aligned}
$$

This proof consists of six steps, each one using a rule of semantic equivalence. The proof illustrates an approach that is central to all of Mathematics and especially to Algebra, *axiomatic proofs*. In an axiomatic proof, certain theorems (called

---

[10] Since we did not formally define the semantics of predicate logic in this chapter, we should rather speak of this as an example of an *intuitively semantic* proof. In Section 4.3 we will produce more formal semantic proofs.

axioms, the ones stated in Lemma 2.1 for the above example proof) are given and
assumed to be true. And then one may use proof rules (that are known to be valid)
to derive the validity of a claim from the assumed axioms. The choice of axioms
and proof rules will depend on what we claim and what we need to derive from
that claim. For example, we may only have proof rules for equality, and so these
rules capture that we may replace equals for equals. An axiom may model a certain domain within which we reason. For example, we may need an axiom such as
$\forall n. \, n + 0 = 0$ when deriving claims about the natural numbers.

> **Semantic equivalence** The meaning of "semantic equivalence" depends on
> the chosen logic and even on the structures that this logic refers to: we already
> defined semantic equivalence for propositional logic through equality of truth
> tables. For predicate logic, we cannot use this definition since atomic propo-
> sitions are evaluated over structures. Intuitively, we say that formulas $\mathscr{A}$ and
> $\mathscr{B}$ of predicate logic are semantically equivalent if they are true in the same
> structures. So, although the formula $\exists x. \, \forall y. \, (x = y) \vee (x < y)$ is equivalent
> to $\forall y. \, (0 = y) \vee (0 < y)$ over the natural numbers $\mathbb{N}$, these formulas are not
> semantically equivalent. We leave it as an exercise to find a structure with an
> interpretation of 0 and $<$ in which one of these formulas is true, and the other
> one is false.
>
> Since the meaning of propositional operators in formulas of predicate logic
> is that of propositional logic, we obtain that semantic equivalences for propo-
> sitional logic are also semantic equivalences for predicate logic. For example,
> the semantic equivalence $\mathscr{A} \Rightarrow \mathscr{B} \equiv \neg \mathscr{A} \vee \mathscr{B}$ of propositional logic also holds
> for all formulas $\mathscr{A}$ and $\mathscr{B}$ of predicate logic.

Let us now give a more formal account of the proof rules needed for axiomatic
reasoning based on semantic equivalences. These proof rules are known as the *Theory of Equational Reasoning* and are shown in Figure 2.1. These rules operate at the
meta-level and therefore are not rules within the logic itself. The first three proof
rules just codify our intuition about equality: everything is equal to itself, equality
between two things is not dependent on the order in which we compare them, and
all things on a chain of pairwise equal links have to be equal.

1. **Reflexivity:** $\mathscr{A} \equiv \mathscr{A}$
2. **Symmetry:** $\mathscr{A} \equiv \mathscr{B}$ implies $\mathscr{B} \equiv \mathscr{A}$
3. **Transitivity:** $\mathscr{A} \equiv \mathscr{B}$ and $\mathscr{B} \equiv \mathscr{C}$ imply $\mathscr{A} \equiv \mathscr{C}$
4. **Compositionality:** $\mathscr{A} \equiv \mathscr{A}'$ implies $\mathscr{B}(\mathscr{A}) \equiv \mathscr{B}[\mathscr{A} \mapsto \mathscr{A}']$

Fig. 2.1 Proof rules for equational reasoning, where $\mathscr{B}(\mathscr{A})$ denotes a formula $\mathscr{B}$ that contains one
or more occurrences of formula $\mathscr{A}$, and $\mathscr{B}[\mathscr{A} \mapsto \mathscr{A}']$ denotes the formula obtained by replacing
all occurrences of $\mathscr{A}$ in $\mathscr{B}(\mathscr{A})$ with $\mathscr{A}'$

The last proof rule is the most interesting one, and captures that we may replace equals for equals *in context*. This rule uses a *syntactic substitution* of one

sub-formula with another in a given formula. Syntactic substitution is an important concept in Informatics that we will revisit in Chapter 4 (cf. Definition 4.6).

In the above proof, for example, we inferred the semantic equivalence $\top \vee \mathscr{A} \equiv (\mathscr{A} \vee \neg \mathscr{A}) \vee \mathscr{A}$ from semantic equivalence Negation by an application of this proof rule. Here we replace $\top$ in $\top \vee \mathscr{A}$ with $(\mathscr{A} \vee \neg \mathscr{A})$ to obtain $(\mathscr{A} \vee \neg \mathscr{A}) \vee \mathscr{A}$. It should be intuitively clear that the proof rules for Equational Reasoning are valid.

We record this general approach informally in the following proof pattern.

Proof Principle 2 (Axiomatic Proofs)

Based on a set of axioms, that is on theorems that are presupposed *to be true, the claim to be shown will be derived by applying to these axioms proof rules that are known to be valid.*

A good example of axiomatic proofs is where claims are about semantic equivalences, where certain semantic equivalences are given as axioms, and where we may *replace equals with equals* by appealing to such axioms to derive the claimed semantic equivalence; the appeal to the proof rules in Figure 2.1 is often omitted in semi-formal proofs.

> Compared to the manner in which we will formulate proof principles subsequently, the definitions of Proof Principles 1 and 2 were deliberately somewhat vague. But we will subsequently develop Proof Principle 2 further and make it more precise (for example in Section 5.7.3). In particular, the reflections on this chapter, found in Section 2.3.2, will mention numerous supplementary aspects of this proof principle.

> **Informatics can validate mathematical proofs** In Mathematics, it is traditional to use a hybrid of the semantic and axiomatic proof principles, and these proofs are often produced in a semi-formal manner, or in a formal manner that won't be executable on a computer. This informality becomes a concern for some mathematical problems that have very long and complex proofs, even though the statements to be proved may seem very simple. One good example of this is Kepler's Conjecture, which the mathematician and astronomer Johannes Kepler formulated in the seventeenth century. For the following discussion you do not need to know this conjecture nor its details. In simplified terms, it says that stacking oranges in a pyramid shape is the most space efficient way of packing oranges.
>
> At the time, the problem was formulated with cannonballs instead. Kepler's Conjecture seems pretty simple as a statement. But it took a long time to prove it. In 1998 Thomas Hales claimed to have found such a proof [30]. But the proof was so long and complex that mathematical peers could not vouch

for its correctness with a high degree of confidence. This motivated Hales to reproduce these proofs in an interactive theorem prover [57] so that the proofs could be type checked in a proof language that has a small set of axioms and inference rules. The so-called Flyspeck Project managed to complete this task in 2014, which also resulted in simplifying the original proof. The degree of confidence in the Flyspeck proof is now much higher, as we only have to trust the implementation of the few axioms and inference rules in these theorem provers.

This discussion also illustrates that theorems and proofs, although they are mathematical objects of great precision, are also social constructs that reflect general human behavior, beliefs, and limitations.

Compositionality. Proofs based on truth tables establish results directly at the semantic level through systematic enumeration of all possibilities (see the discussion in Section 4.3). In contrast, axiomatic proofs operate at the syntactic level. They solely rely on the transformation of concrete (syntactic) representations, where these transformations preserve the semantics of the transformed syntactic object. This preservation of meaning by single transformation steps is rooted in the validity of all axioms and (if needed) in the validity of theorems and lemmas that can be shown to be logical consequences of these axioms.

The correctness of this overall approach rests on the principle of *compositionality* of the underlying set of axioms. You will already know this principle from school, in the form of the algebraic principle of *Substituting Equals for Equals* in the manipulation of algebraic terms and equations. The principle of compositionality here means that the semantics of a complex construct does not depend on representational aspects of its components, but only on the semantics of those components. To illustrate, we can compute the result s of the addition $a + b$ by knowing what a and b have as value – regardless of how the values of a and b might have been determined (for example a itself might be the sum or product of other numbers). The proof rules in Figure 2.1 capture this compositionality well.

Compositionality is a leading paradigm in Informatics: it is a guiding principle for defining the meaning of caluli, logics, and programming languages – to name a few. Sometimes it can be challenging to attain compositionality. Let us discuss an example from propositional logic: we will call a formula $\mathcal{A}$ of propositional logic *satisfiable* iff there is an assignment that makes $\mathcal{A}$ true. For example $\mathcal{A} \lor \neg \mathcal{A}$ is satisfiable whereas $\mathcal{A} \land \neg \mathcal{A}$ is not. In the exercises, you should show that atomic propositions and their negations are satisfiable; and that the logical disjunction $\mathcal{A} \lor \mathcal{B}$ is satisfiable if one of $\mathcal{A}$ or $\mathcal{B}$ is satisfiable already. So this may suggest that there is a compositional way of computing which formulas are satisfiable. But what about the logical operators $\neg$ and $\land$? For these, compositionality of "satisfiability" breaks down: for example, in the exercises you are asked to find formulas $\mathcal{A}$ and $\mathcal{B}$ such that $\mathcal{A}$ and $\mathcal{B}$ are satisfiable whereas neither $\neg \mathcal{A}$ nor $\mathcal{A} \land \mathcal{B}$ are satisfiable.

The difficulty of maintaining compositional reasoning can also be seen in the design of programming constructs that support parallel execution based on threads or processes. There is an interesting friction in Informatics between thinking of reasoning support as something to add to a design or implementation a posteriori, or as something that should inform decisions already at design stage. You can see this also in the area of parallel computing, where reasoning about mainstream parallel languages is hard and requires a lot of research and tool development, but where there is also a strand of research that tries to invent domain-specific languages for more controlled parallel compositions that better facilitate reasoning [82].

Direct versus Indirect Proofs The proof principles we saw so far – be they semantic, axiomatic, or a combination thereof – all shared that they aimed to prove a claim directly. These approaches therefore abide by the principle of *Direct Proof*. However, this approach will not always work or be effective. For example, consider Euclid's Theorem – which states that there are infinitely many primes. A direct proof of this would have to somehow construct infinitely many prime numbers, presumably by finding some parameterized formula for generating infinitely many primes. But the distribution of prime numbers in $\mathbb{N}$ seems to make such an approach very hard. In this case, it is much simpler to prove this *indirectly* by assuming the negation of the theorem to be true, i.e., we then assume that there are only finitely many prime numbers $p_1, \ldots, p_k$ in $\mathbb{N}$. Then we show that this assumption leads to a contradiction.

Consider the product p of all these prime numbers p_i, i.e., $p =_{df} p_1 \cdot p_2 \cdot \ldots \cdot p_k$. But then $p + 1$ must also be a prime number: if $p + 1$ has a divisor other than 1 and $p + 1$, then it also has such a divisor q that is a prime number. But then q would have to equal p_i for some i as these are claimed to be all the prime numbers. This yields a contradiction as q certainly divides p, but not 1 and therefore, by distributivity, $p + 1$ cannot have q as a divisor. From this contradiction, we now infer that the initial assumption, that there are only finitely many primes, must be wrong. And this proves Euclid's Theorem.

The principle on which the former proof rests is known as *Proof by Contradiction* or also *reductio ad absurdum*.

> **Proof Principle 3 (Proof by Contradiction)**
>
> ---
>
> *Let $\mathscr{A}$ be a proposition that we wish to prove. Suppose that under the assumption that $\neg\mathscr{A}$ holds we are able to derive falsity F, for instance by proving that a proposition $\mathscr{B}$ together with its negation $\neg\mathscr{B}$ must hold. Then it must be the case that proposition $\mathscr{A}$ holds.*
>
> *Expressed in symbolic reasoning, this means that the formula*
>
> $$(\neg\mathscr{A} \Rightarrow \mathsf{F}) \Rightarrow \mathscr{A}$$
>
> *is always valid.*

The correctness of this proof principle can be established through an inspection of the corresponding truth tables, which we leave as an exercise. In Chapter 3 we will encounter another important proof principle for indirect proofs: *Proof by Contraposition* (see Proof Principle 4).

Law of the Excluded Middle The expressive power of indirect proofs rests on the *Law of the Excluded Middle*, known already as *tertium non datur* in Latin. We can explain this law semantically: there are only two truth values, *true* and *false*, not any third value. For example, there either are infinitely many primes or there are not (in which case there are then only finitely many primes). This Law of the Excluded Middle is criticized or abandoned by some Mathematicians, Philosophers, and Informaticians. Indeed, one can develop *constructive, intuitionistic logics* for the development of *constructive mathematics* [54], where proofs always render an explicit construction that explains the validity of the proved claim.

> To illustrate the "disturbing" power of the Law of the Excluded Middle, let us consider the claim *"There are irrational numbers a and b such that a^b is rational."* We can prove this claim indirectly. Let
>
> $$c =_{df} \sqrt{3}^{\sqrt{2}}$$
>
> This number is either rational or irrational (there is no third possibility). If c is rational, we may choose $a =_{df} \sqrt{3}$ and $b =_{df} \sqrt{2}$. Otherwise, c is irrational and we may choose $a =_{df} c$ and $b =_{df} \sqrt{2}$. You should check that in both cases it follows that a^b is now rational.
>
> But does this indirect proof tell us values of irrational numbers a and b such that a^b is rational? It does not, it only shows that it is always possible to find such values. Constructive Mathematicians and constructive Logicians would therefore refuse to accept such a proof. Although such a refusal may seem dogmatic it actually leads to many interesting logics that have important applications in Informatics. This hints at a crucial difference in the role of logic in Mathematics and Informatics. For Mathematics, logic is a more or

less fixed tool that gets used every day. In Informatics, logic is a toolbox and we can and often do *engineer* a new logic to solve a problem in Informatics for us.

2.2 Sets

Set Theory, as an independent area of mathematical study, can be traced back to the work of Georg Cantor in the late nineteenth century. In the course of his work, he offered the following definition of sets.

Definition 2.4 (Set). A *set* is understood to be a collection S of certain, well-distinguished objects of our thought or considerations into a whole. These objects are called the *elements* of S.   □

For a set S, the proposition $m \in S$ means that m is an element of set S. The negation of this proposition, $\neg(m \in S)$, is often abbreviated by $m \notin S$.

We can describe sets in different ways. Sets that contain only finitely many elements, called *finite sets*, can be specified by an *enumeration* of all their elements. For example,

$$S =_{df} \{\clubsuit, \spadesuit, \heartsuit, \diamondsuit\}$$

is the set of all suits in a four-suit deck of cards, whereas

$$W =_{df} \{\text{Monday, Tuesday, Wednesday, Thursday, Friday, Saturday, Sunday}\}$$

is the set of days of the week in the Gregorian Calendar.

We can "enumerate" sets with infinitely many elements as well, provided that the elements follow a clear pattern that this enumeration can indicate. For example, we can specify the set of natural numbers $\mathbb{N}$, familiar from school, by the definition $\mathbb{N} = \{0, 1, 2, 3, \ldots\}$ and the set of even natural numbers can be defined as $\mathbb{N}_{ev} = \{0, 2, 4, 6, \ldots\}$.[11]

The ...-notation seems to be convenient, but it is not unproblematic. In the above examples most of us would instantly recognize the patterns at work in these enumerations, and so would be able to expand the explicit enumeration of those sets correctly to any length desired. However, consider the enumeration of the set $P = \{303, 2212, 7549, 18102, \ldots\}$. What should we make of this definition? It turns out that the first four elements enumerated in set P correspond to the values $p(i)$ for $i = 1, 2, 3, 4$ for the polynomial $p(x) =_{df} 34 + 45 \cdot x - 74 \cdot x^2 + 298 \cdot x^3$. Although this seems to suggest that we have found a pattern, there is no guarantee that the fifth element in set P

---

[11] In Chapter 4 we will introduce the concept of *inductive thinking* as a formalization of the meaning of the "..."-notation.

will equal $p(5)$, i.e., $35,659$; set P may well be based on another enumeration pattern! The twentieth century philosopher Ludwig Wittgenstein was already aware of such issues in the pattern-based description of infinite sets. We learn from this that the ...-notation is limited and needs a more formal and unambiguous equivalent. In fact, this is particularly important for Informatics, e.g., to obtain precise and concise definitions of the syntax of programming languages (cf. Section 4.3.3).

There is also a *descriptional* way of defining sets, and this is known as the *intensional* characterization of sets or as *set comprehension*. The way this works is that the elements contained in a set are characterized by a predicate (or formula more generally), and the set therefore consists of all and only those elements that satisfy the predicate. The form for such definitions is

$$S = \{m \mid \mathscr{A}(m)\}$$

where $\mathscr{A}(m)$ is a formula of predicate logic that has m as a free variable, and S is the set defined by this formula.

For example, the set of prime numbers can then be defined by[12]

$$Primes \ =_{df} \ \{p \mid p \in \mathbb{N} \wedge p \neq 1 \wedge \forall n \in \mathbb{N}.\, n \mid p \ \Rightarrow \ n = 1 \vee n = p\} \qquad (2.6)$$

The formula used here is $\mathscr{A}(p) \ =_{df} \ p \in \mathbb{N} \wedge p \neq 1 \wedge \forall n \in \mathbb{N}.\, n \mid p \ \Rightarrow \ n = 1 \vee n = p$ and this does have p as (only) free variable.

In case that all elements of $S =_{df} \{m \mid \mathscr{A}(m)\}$ are from some known set T, we may write a variant form of set comprehension as $S =_{df} \{m \in T \mid \mathscr{A}(m)\}$. In the above definition of *Primes*, it is clear that all elements of *Primes* are in $\mathbb{N}$, since $p \in \mathbb{N}$ is a conjunct of $\mathscr{A}(p)$. Therefore, we may define *Primes* also more compactly by

$$Primes =_{df} \{p \in \mathbb{N} \mid p \neq 1 \wedge \forall n \in \mathbb{N}.\, n \mid p \ \Rightarrow \ n = 1 \vee n = p\}$$

An interesting boundary case of set comprehension is when the formula $\mathscr{A}(m)$ is always false, a property we will later call *unsatisfiable*. Then $\{m \mid \mathscr{A}(m)\}$ defines the *empty set*, written $\emptyset$, which is distinguished by the property of not having any elements. In particular, we have $\emptyset = \{x \mid \mathsf{F}\}$ and the formula $\forall x.\, x \notin \emptyset$ is always true. It is perhaps surprising that the formula $\forall x \in \emptyset.\, \mathscr{A}(x)$ is also always true. The reason is that the qualified quantification $\forall x \in \emptyset.\, \mathscr{A}(x)$ is just a convenience for writing $\forall x.\, x \in \emptyset \Rightarrow \mathscr{A}(x)$. And the latter formula is always true since $x \in \emptyset$ is always false.

That $\forall x \in \emptyset.\, \mathscr{A}(x)$ is always true is a variant of a proof principle called *ex falso quodlibet* (from falsehood we may infer whatever we like). This principle not only

---

[12] This definition of the set of prime numbers *Primes* and an alternative definition thereof given below suffer from an unfortunate overloading of the symbol "$\mid$": it is used in the set comprehension form $S = \{m \mid \mathscr{A}(m)\}$ but it also denotes the divisibility relation $n \mid m$. Section 4.5.1 discusses such, in a certain way unavoidable, notational overloading. In this trilogy, we will occasionally encounter such notational ambiguities, but context will always allow us to resolve these ambiguities correctly.

alienates many of those who are unfamiliar with formal mathematics, it is also a common source of erroneous reasoning in what may otherwise seem like conclusive proofs: falsehood typically only occurs as temporary evidence in indirect proofs, unless the axioms or inference rules themselves are invalid.

We will discuss sets of sets and set equality below. But let us point out here already that the empty set $\emptyset$ is different from the set $\{\emptyset\}$. The former contains no elements, whereas the latter contains exactly one element: the empty set $\emptyset$.

All sets have two intrinsic properties:

- Elements of a set do not have a notion of frequency. For example, the sets $\{1,2\}$ and $\{2,1,2\}$ are equal. Sometimes, it is convenient to account for such frequencies and to then distinguish sets not just in terms of their elements but also in terms of how often these elements occur: this leads to the development of so-called *multi-sets*.
- A set does *not* define an order of its elements. For example, the sets $\{1,2,3\}$ and $\{3,1,2\}$ are equal.[13]

2.2.1 Set Relationships

In the consideration of sets, it is often of interest to understand relationships between them. A natural question in this context is, of course, whether two sets contain the same elements, which by virtue of Cantor's definition of sets asks whether the two sets are equal. But even sets that are not equal can have interesting relationships. For example, if all elements of one set are also elements of another set, we have a *Subset Relationship*, also called a *Set Inclusion*. We may capture these and other important set relationships in the following formal definitions.

Definition 2.5 (Set relationships). Let A and B be sets. We define the following relationships:

1. $A \subseteq B$ (pronounced "A is a *subset* of B") $\Leftrightarrow_{df}$ $(\forall x. \, x \in A \Rightarrow x \in B)$

2. $A = B$ (pronounced "A *equals* B") $\Leftrightarrow_{df}$ $A \subseteq B \wedge B \subseteq A$

3. $A \subset B$ (pronounced "A is a *proper subset* of B") $\Leftrightarrow_{df}$ $A \subseteq B \wedge A \neq B$.[14]   □

By definition of the empty set, it is a subset of all other sets. The (proper) subset relationship of sets can be graphically illustrated in so-called *Venn diagrams* (see Figure 2.2). In a Venn diagram, sets are represented by geometric circles (and their interior). Relationships between sets are represented by appropriately overlapping (i.e., positioning) such circles. For example, two circles that don't overlap model

---

[13] We will define and discuss orders that are important in Informatics in the second volume, *Algebraic Thinking*.

[14] $A \neq B$ denotes here $\neg(A = B)$

that the two sets represented by these circles have no element in common. And a circle fully contained in another circle represents a subset relationship.

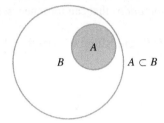

Fig. 2.2 Venn diagram for proper subset relationship

2.2.2 Power Sets

Sets whose elements are sets themselves are called *families of sets* or *families of sets over S* if all sets in that family are from the same ground set S. A very prominent example of families of sets is power sets, which contain *all* subsets of some ground set S.

Definition 2.6 (Power set). Let S be a set. The *power set* of S is defined by $\mathfrak{P}(S) =_{df} \{S' \mid S' \subseteq S\}$.   □

Note how this definition makes use of the predicate $A \subseteq B$ we defined above. It is worth pointing out that all power sets $\mathfrak{P}(S)$ contain S and the empty set $\emptyset$, since we always have $S \subseteq S$ and $\emptyset \subseteq S$. A special case is that of $S = \emptyset$. In that case, S and $\emptyset$ are the same sets and the only element in $\mathfrak{P}(S)$. In particular, and perhaps surprisingly to you, the power set of the empty set is not empty! In Section 3.2.3, we will formally prove that the power set of an *arbitrary* (even infinite) set is *strictly larger* than the set itself. This will require a definition and understanding of the *size* of sets, be they finite or infinite. And this result will mean that there is an infinite hierarchy of infinities in set theory!

Example 2.4. For the set of ground elements $S = \{1, 2, 3\}$ we have

$$\mathfrak{P}(S) = \{\emptyset, \{1\}, \{2\}, \{3\}, \{1, 2\}, \{1, 3\}, \{2, 3\}, \{1, 2, 3\}\}$$

Power sets form a mathematical structure called a *complete lattice*, where the elements of the power set are ordered with respect to subset inclusion. Such structures are of importance in Informatics, in particular in the *Semantics of Programming Languages* [56, 66]. The second volume of this trilogy, *Algebraic Thinking*, will discuss the foundational theory of complete lattices and will explain why power sets are a prototype of complete lattices such that the latter are an abstraction of the former that preserves essential properties.

2.2.3 Composing Sets

Similarly to how logical operators compose formulas into more complex formulas, there are *set operations* that can form new sets from existing ones.

Definition 2.7 (Composition of sets). Let A and B be sets. Then we define the following set operations:

- *Union* $A \cup B =_{df} \{x \mid x \in A \lor x \in B\}$

- *Intersection* $A \cap B =_{df} \{x \mid x \in A \land x \in B\}$

- *Difference* $A \setminus B =_{df} \{x \mid x \in A \land x \notin B\}$

- *Symmetric Difference* $A \Delta B =_{df} (A \cup B) \setminus (A \cap B)$ □

Sometimes, the names of these operations also contain the adjective "set-theoretic" as in "set-theoretic union" to stress that these operations are meant to apply to sets. The above definitions reveal that the logical operators $\land$ and $\lor$ share more with the set-theoretic operators $\cap$ and $\cup$ (respectively) than superficial similarity. Rather, we will see that the former are essential for defining the meaning of the latter. Moreover, the symmetric difference operator Δ corresponds to the logical XOR operator since the sets $A \Delta B$ and $\{x \mid x \in A \oplus x \in B\}$ can be shown to be equal.

These connections can be made more precise as follows. Consider a relationship $\leq$ and an operation $\sqcap$ with the following set of axioms:

$$x \leq x \tag{2.7}$$
$$x \sqcap y \leq x$$
$$x \sqcap y \leq y$$
$$(z \leq x) \land (z \leq y) \text{ implies } (z \leq x \sqcap y)$$
$$(x \leq y) \land (y \leq x) \text{ implies } (x = y)$$

In our discussion of lattices in the second volume, *Algebraic Thinking*, we will see that $\sqcap$ defines the "meet" operation in a lattice with partial order $\leq$. The first axiom says that each element is less than or equal to itself. The meet $x \sqcap y$ is below both x and y in the order (the next two axioms); and any element that is below x and y in the order is also below the meet of x and y in the order (fourth axiom). The fifth axiom says that elements are equal if they are below each other.

What is remarkable is that these axioms hold in set theory and in propositional logic; in the former, we interpret $\leq$ to mean subset inclusion $\subseteq$ and $\sqcap$ to mean the intersection of sets $\cap$; in the latter, $\leq$ is interpreted as implication $\Rightarrow$ and $\sqcap$ as logical conjunction $\land$. It is an exercise to show that the axioms above hold under these interpretations.

It is also an exercise to show that the axioms in (2.7) (which all hold in a lattice) allow us to prove that $x \sqcap x = x$. Therefore, we get the set-theoretic law $A \cap A = A$ and the semantic equivalence $\mathscr{A} \wedge \mathscr{A} \equiv \mathscr{A}$ for free from this reasoning at the level of lattices.

Whenever we have that $A \cap B$ equals $\emptyset$, we refer to A and B as being *disjoint*. For a given set S of ground elements, we can define the *complement* of any subset $A \subseteq S$ as $A^C =_{df} S \backslash A$. Note that the notation A^C does not make explicit the ground set over which the complement is taken, so this needs to be understood from context.

Figure 2.3 shows the Venn diagrams of all of the above set operations. The sets that result from the application of these set operations are shown in grey in the figure.

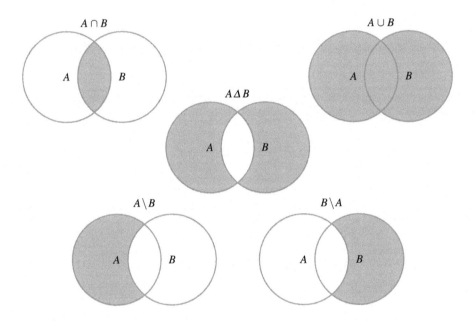

Fig. 2.3 Venn diagrams of set operations

Example 2.5. Let *Primes* and $\mathbb{N}_{ev}$ be the set of prime numbers and the set of even, natural numbers (respectively), as defined on page 43. Then we have the following:

- $Primes \cap \mathbb{N}_{ev} \;=\; \{2\}$

- $Primes \cup \mathbb{N}_{ev} \;=\; \{0, 2, 3, 4, 5, 6, 7, 8, 10, \ldots\}$ Note: $9 \notin Primes \cup \mathbb{N}_{ev}$

- $Primes \backslash \mathbb{N}_{ev} \;=\; \{3, 5, 7, 11, 13, \ldots\}$

- $\mathbb{N}_{ev} \setminus Primes \quad = \{0,4,6,8,10,\ldots\}$
- $\mathbb{N}_{ev} \, \Delta \, Primes \quad = \{0,3,4,5,6,7,8,10,\ldots\}$

Corresponding to the laws of propositional logic, the semantic equivalences expressed in Lemma 2.1, we have the following Equational Laws for sets.

Lemma 2.3 (Equational laws of set operations). *Let A, B, C be subsets of a common set S of ground elements. Then we have:*[15]

$$A \cap B = B \cap A \qquad\qquad\qquad \textit{(Commutativity)}$$
$$A \cup B = B \cup A$$

$$(A \cap B) \cap C = A \cap (B \cap C) \qquad\qquad \textit{(Associativity)}$$
$$(A \cup B) \cup C = A \cup (B \cup C)$$

$$A \cap (A \cup B) = A \qquad\qquad\qquad \textit{(Absorption)}$$
$$A \cup (A \cap B) = A$$

$$A \cap (B \cup C) = (A \cap B) \cup (A \cap C) \qquad \textit{(Distributivity)}$$
$$A \cup (B \cap C) = (A \cup B) \cap (A \cup C)$$

$$A \cap A^C = \emptyset \qquad\qquad\qquad\qquad \textit{(Complement)}$$
$$A \cup A^C = S$$

$$A \cap A = A \qquad\qquad\qquad\qquad \textit{(Idempotency)}$$
$$A \cup A = A$$

$$A^{CC} = A \qquad\qquad\qquad\qquad \textit{(Double Complement)}$$

$$(A \cap B)^C = A^C \cup B^C \qquad\qquad \textit{(De Morgan's Laws)}$$
$$(A \cup B)^C = A^C \cap B^C$$

$$S \cap A = A \qquad\qquad\qquad\qquad \textit{(Neutrality)}$$
$$\emptyset \cup A = A$$

We can generalize the set operations *union* and *intersection* from sets to arbitrary families of sets.

Definition 2.8 (Extended unions and intersections). Let $\mathfrak{S}$ be a family of sets over a set S of ground elements. Then we have:

1. $\displaystyle\bigcup_{S' \in \mathfrak{S}} S' =_{df} \{m \in S \mid \exists S' \in \mathfrak{S}.\ m \in S'\}$

2. $\displaystyle\bigcap_{S' \in \mathfrak{S}} S' =_{df} \{m \in S \mid \forall S' \in \mathfrak{S}.\ m \in S'\}$

We may abbreviate these operations to $\bigcup \mathfrak{S}$ and $\bigcap \mathfrak{S}$ (respectively). □

---

[15] The restriction of having a common set of ground elements applies only to equational laws that refer to either S or any set complements using the operation ^{C}.

The transfer of the concept of *disjoint sets* to families of sets requires some care: suppose that $\bigcap \mathfrak{S}$ equals $\emptyset$. Then this does not at all imply that all the sets in $\mathfrak{S}$ contain elements different from all elements of all other sets in $\mathfrak{S}$. For example, let $\mathfrak{S}$ contain the sets $\{1,2,3\}$, $\{2,3,4\}$, $\{3,4,5\}$, and $\{4,5,6\}$. Then $\bigcap \mathfrak{S}$ equals $\emptyset$ but only the sets $\{1,2,3\}$ and $\{4,5,6\}$ are disjoint. Can you find an example of $\mathfrak{S}$ such that $\bigcap \mathfrak{S}$ equals $\emptyset$ but for all A and B in $\mathfrak{S}$ we have $A \cap B \neq \emptyset$? We leave this question as an exercise. We say that $\mathfrak{S}$ is *pairwise disjoint* if every element of $\bigcup \mathfrak{S}$ is contained in exactly one of the elements of $\mathfrak{S}$. We may express this concept of mutual disjointness in a formula of predicate logic:

$$\forall S_1, S_2 \in \mathfrak{S}. \ (S_1 = S_2) \lor (S_1 \cap S_2 = \emptyset)$$

2.2.4 Cardinality of Finite Sets

Intuitively, a set S is finite if it contains only finitely many elements. So $\{1,2,\ldots,12\}$ is a finite set, whereas $\mathbb{N}$ is not. Note that the set $\{-1, \mathbb{N}\}$ is finite as well since it contains two elements. The fact that the second element is a set (and an infinite one at that) is immaterial here as the elements of $\mathbb{N}$ are not elements of $\{-1, \mathbb{N}\}$.

For a finite set S, we call the number of its elements its *cardinality*, and denote this number by $|S|$.[16] The cardinality of the empty set is 0, and the empty set is the only set with that cardinality: $|\emptyset| = 0$.

Theorem 2.2. *Let A and B be finite sets. Then we have the following Arithmetic Laws:*

1. $|A \backslash B| = |A| - |A \cap B|$

2. $|A \cup B| = |A| + |B| - |A \cap B|$

3. $|A \Delta B| = |A| + |B| - 2 \cdot |A \cap B|$

Proof. 1. Given $C \subseteq A$, we have $|A \backslash C| = |A| - |C|$, since

$$|A \backslash C| = |\{a \in A \mid a \notin C\}| = |A| - |\{a \in A \mid a \in C\}| = |A| - |C| \quad (2.8)$$

Assume, for now, that we have $A \backslash B = A \backslash (A \cap B)$. Setting $C =_{df} A \cap B$, we then have $C \subseteq A$ and so the above equation (**??**) means that

$$|A \backslash B| = |A \backslash (A \cap B)| = |A| - |A \cap B|$$

Therefore, it suffices to show that the assumption $A \backslash B = A \backslash (A \cap B)$ is true. By Definition 2.5(2), we can show this set equality by showing two set inclusions.[17]

---

[16] Sometimes, one also encounters the notation $\sharp S$, for example in order to disambiguate this notation from that for absolute values $|\cdot|$ of real numbers.

[17] The definition of set equality through set inclusions may seem non-standard but is that of Axiomatic Set Theory. Proofs of set equality often appeal to that definition directly (as in this case).

"⊆": Let a be in $A\backslash B$. Then we have $a \in A$ and $a \notin B$. In particular, the latter implies $a \notin A \cap B$ and therefore also $a \in A\backslash(A \cap B)$.

"⊇": Let a be in $A\backslash(A \cap B)$. Then we have $a \in A$ and $a \notin A \cap B$. From the latter and the De Morgan's Law it follows that $a \notin A$ or $a \notin B$. Together with $a \in A$ we get $a \notin B$ and so also $a \in A\backslash B$.

2. For a finite set C that is disjoint to (finite) set A we have clearly that

$$|A \cup C| = |A| + |C| \qquad (2.9)$$

Furthermore, we have that $A \cup B = A \cup (B\backslash A)$, since

$$A \cup (B\backslash A) = \{a \mid a \in A \vee (a \in B \wedge a \notin A)\}$$
$$\overset{(\text{Distr.})}{=} \{a \mid (a \in A \vee a \in B) \wedge (a \in A \vee a \notin A)\}$$
$$\overset{(\text{Neg.})}{=} \{a \mid (a \in A \vee a \in B) \wedge \mathsf{T}\}$$
$$\overset{(\text{Neutr.})}{=} \{a \mid a \in A \vee a \in B\} = A \cup B$$

By definition, $B\backslash A$ is disjoint to A. Setting now $C =_{df} B\backslash A$, we can apply the above equation (2.9) to obtain

$$|A \cup B| = |A \cup (B\backslash A)| = |A| + |B\backslash A| \overset{\text{Part 1}}{=} |A| + |B| - |A \cap B|$$

3. Because we have $A \Delta B = (A \cup B)\backslash(A \cap B)$, we also get

$$|A \Delta B| \overset{\text{Part 1}}{=} |A \cup B| - |A \cap B| \overset{Part2}{=} |A| + |B| - 2 \cdot |A \cap B|$$

$$\square$$

It is not apparent how to generalize the concept of cardinality from finite to infinite sets. In fact, we don't yet have a formal definition of when a given set is finite. In Section 3.2.3, we will see that there are even different kinds of infinities when counting the number of elements of infinite sets! These insights will be elegantly explained by a formalization of the concept of cardinality.

Antinomies The so-called "Cantorian Set Theory", which we presented here so far, reaches its limits in the naive and uncontrolled application of its concept of sets and set operations. In this naive set theory, it is possible to define sets that, upon closer inspection, contain or lead to logical contradictions. Such paradoxical sets are known as *antinomies* and will be further discussed in our more advanced reflections in Section 2.3.5.

We will generalize this proof principle to the so-called Anti-Symmetry Proof Principle and will prove the underlying Theorem of Schröder–Cantor–Bernstein in the second volume of this trilogy.

2.3 Reflections: Exploring Covered Topics More Deeply

This and all subsequent chapters will be concluded with reflections on the developed material, where these reflections deepen the understanding of the covered topics.

2.3.1 Propositional Logic

Size of truth tables In the study of formulas of propositional logic, we are rarely interested in an explicit account of truth values for all possible assignments of truth values to propositional atoms. Rather, the following questions are often of immediate interest:

- Is a formula of propositional logic *inconsistent* (also known as *unsatisfiable*), i.e., is there no assignment that makes the formula true?

- Is a formula *consistent* (also known as *satisfiable*), i.e., is there an assignment that makes the formula true?

- Or is the formula even a *tautology* (also known as *valid*), i.e., is the formula true under all assignments?

For example, for arbitrary atomic propositions $\mathscr{A}$ and $\mathscr{B}$, the formula $\mathscr{A} \vee \mathscr{B}$ is satisfiable and not a tautology. The formula $(\mathscr{A} \Rightarrow \mathscr{B}) \vee (\mathscr{B} \Rightarrow \mathscr{A})$ is a tautology, and $\mathscr{A} \wedge \neg \mathscr{A}$ is unsatisfiable.

You can quickly see that it won't always be necessary to construct the entire truth table of a formula to answer these questions unambiguously. For example, to show that a formula is satisfiable, it suffices to find one way of assigning truth values to atomic propositions that makes the formula true. Similarly, to show that the formula is *not* a tautology, it suffices to find one way of assigning truth values to atomic propositions for which the formula is false. However, suppose that a formula is valid. Then any attempt to demonstrate this by computing truth values for assignments of truth values to atomic propositions will necessarily have to construct the entire truth table for that formula, for example in the case of tautology $(\mathscr{A} \Rightarrow \mathscr{B}) \vee (\mathscr{B} \Rightarrow \mathscr{A})$.

We have already remarked that the size of the truth table grows exponentially with the number of atomic propositions in the formula in question: if there are k such atomic propositions, then the truth table has 2^k rows. For example, for two atomic propositions, there are four rows and these are easy to inspect; for five atomic propositions, there are 32 rows and it is already a bit harder to construct this many rows by hand without any errors; for 50 atomic propositions, we can no longer construct truth tables manually, and even computers will struggle to construct 2^{50} rows. To illustrate the sheer size of this, 2^{50} equals 1,125,899,906,842,624.

In Informatics, there are specialized algorithms that can often decide whether
a formula is satisfiable or unsatisfiable; and these algorithms won't construct truth
tables and will have sophisticated search and inference methods that aim to minimize the number of cases to consider when answering such questions. For example,
modern so-called *Boolean Satisfiability Solvers*[18] [28] (often referred to as "SAT
solvers") can answer such questions for many formulas that have several hundreds
or even thousands of atomic propositions. In particular, these algorithms terminate
by either reporting an assignment that makes the input formula true (also known as
a "satisfiability witness") or reporting that the input formula is unsatisfiable.

Application: Digital Circuits In propositional logic, the focus is on the combination of atomic propositions to form more complex ones. The internal structure of
atomic propositions is not relevant in these compositions. The use of propositional
logic is therefore appropriate when base objects are not closely inspected and can
only take on two values. For example, this is the case in the theory of digital circuits.
Such circuits process binary signals which are represented by different voltage levels (also called "states"). For example, a level of 0 volts may represent binary value
0 and corresponds to *ff*; whereas a level of 5 volts may represent binary value 1 and
so correspond to value *tt*.

Let us first consider here the functionality of a so-called half adder. This function
adds two binary values obtained from input wires A and B, whose input signals are
either 0 or 1, to get their sum S and produces a carry C in case that both input signals
have value 1. Such carries may be used by subsequent full adders, for example to add
two natural numbers represented in binary. The logic of this function is as follows:
if the input wires A and B both have input value 1, then S is set to 0 and C is set to
1. In all other cases, S is set to the sum of the input values of A and B, and C is set
to 0.

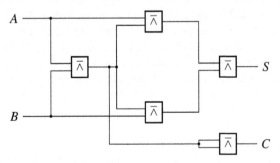

Fig. 2.4 Half adder built out of NAND gates

In Figure 2.4, we see a circuit diagram of a half adder that is exclusively built out of
NAND gates. We know that this is possible since all functions of propositional logic
can be generated with the logical NAND operator (recall the discussion on page 26).
The truth tables that specify the behavior of a half adder are found in Table 2.3.

---

[18] See http://en.wikipedia.org/wiki/Boolean_satisfiability_problem

A	B	D $\overbrace{A \barwedge B}$	E $\overbrace{A \barwedge D}$	F $\overbrace{D \barwedge B}$	S $\overbrace{E \barwedge F}$	C $\overbrace{D \barwedge D}$
0	0	1	1	1	0	0
0	1	1	1	0	1	0
1	0	1	0	1	1	0
1	1	0	1	1	0	1

Table 2.3 Truth tables for the circuit diagram in Figure 2.4. The names D, E, and F are used to denote auxiliary input/output wires

Suppose that the specification of the behavior of a digital circuit can be given by a formula of propositional logic. Then *design synthesis* is understood to be the process of finding a composition of logic gates that realizes exactly the behavior specified in that formula. In that sense, the composition of NAND gates in Figure 2.4 is a correct synthesis of the specification of a half adder. However, practical circuit design needs to consider factors other than mere behavior. For example, there are space constraints to consider in the two-dimensional layout in which logic gates can be burned into silicon. It is also likely that future chips will be more three-dimensional in layout, to exploit parallelism and to allow for an increase in the number of transistors put into an integrated circuit. We won't go much deeper into the topic of digital circuits and refer to [79] for further study of that topic.

2.3.2 Axiomatic Proofs

Central to the concept of an axiom is the fact that an axiom is assumed to be true, without us being under any strict obligation to show that the axiom is indeed true. This is because the meaning of a set of axioms is that they define an abstract world, or rather a set of abstract worlds: all those worlds in which all axioms from that set are true. We can then use inference rules (for example those in Figure 2.1) on these axioms to derive lemmas and theorems. And we can then apply these same inference rules on axioms, derived theorems, or lemmas to derive further theorems and lemmas, and so forth.

Let us suppose that we design all inference rules such that they can only produce true facts from facts that are assumed to be true. Then we know that all theorems and lemmas derived from the set of axioms will be true in *all* abstract worlds that satisfy all axioms from that set. We see that this therefore yields a very powerful reasoning principle. Moreover, this approach is not concerned with how closely these abstract worlds correspond to the real world: it is only concerned with discovering facts about a set of abstract worlds defined by a set of axioms.

To make this more concrete, consider the set of axioms shown in Figure 2.5 below. We can use standard inference rules to show that this set of axioms can derive another theorem: $\forall y. \forall x. (y+x)+x = y$. This means that adding an element x to some element y twice recovers the element y. We said that this derivation is now valid in all abstract worlds that satisfy this set of axioms. So let us get a feel for what some of these abstract worlds might be.

$$\forall x. \, \forall y. \, \forall z. \, x + (y+z) = (x+y) + z$$
$$\forall x. \, \forall y. \, x + y = y + x$$
$$\exists n. \, \forall x. \, (x+n = x) \wedge (x+x = n)$$

Fig. 2.5 Example of a set of axioms

For example, let the structure of individual elements be all bit vectors of length 8, i.e., strings $b_1 b_2 \ldots b_8$ such that all b_i are either 0 or 1. And let $+$ be interpreted as the bitwise exclusive or: for example, $01100110 + 01110111 = 00010001$. Then it is easy to show that all axioms stated in Figure 2.5 are valid for this structure: for the existentially quantified variable n we can choose the witness element 00000000. In particular, we then conclude that the derived theorem $\forall y. \, \forall x. \, (y+x) + x = y$ is also true for this structure.

The set of natural numbers $\mathbb{N}$ with the usual interpretation of $+$ does not satisfy this set of axioms, however. It satisfies the first two, as addition is associative and commutative. But the only natural number n such that $x + n = x$ for all natural numbers is 0; yet it is not true that $x + x$ equals 0 for all x, for example $1 + 1 = 2 \neq 0$. In particular, since the set of axioms is not true for this structure, the derived theorem is not necessarily true in that structure (it is indeed false in this structure).

If we were to interpret $+$ over $\mathbb{N}$ as $x + y =_{df} 0$ for all x and y in $\mathbb{N}$, then the three axioms are now true and so the derived theorem is true as well. This illustrates that the same set of individual elements may or may not satisfy a set of axioms, depending on the interpretation of non-logical operators in those axioms.

An interesting and troublesome boundary case of this approach is when the set of axioms is *inconsistent*. If that set is finite, we may think of such inconsistency as saying that the logical conjunction of all axioms is inconsistent (as discussed on page 52). In that case, there is no abstract world at all in which all these axioms are true, and inferring more facts is therefore a pointless endeavor.

This also shows another application of answering whether a formula (or a set of formulas) of a logic is consistent. Figure 2.6 shows a set of axioms that is inconsistent. The second axiom states that every element is equal to itself. The third axiom says there is an element that is smaller (regardless of how "smaller" will be interpreted in the structure) than all elements. In particular, that element will have to be "smaller" than itself. But then the first axiom implies that this element is not equal to itself, which contradicts the second axiom. Therefore, there cannot be a structure that satisfies all three axioms.

Another nice feature of axiomatic proofs is that they may apply to more than one logic. For example, consider the semantic equivalences above the horizontal line in Lemma 2.1 as an (abstract) set of axioms. We may then show, for example using truth tables and the semantics given in Definition 2.2, that this set of axioms is satis-

$$\forall x.\ \forall y.\ (x < y) \Rightarrow \neg(x = y)$$
$$\forall x.\ x = x$$
$$\exists n.\ \forall x.\ (n < x)$$

Fig. 2.6 Example of an inconsistent set of axioms

fied by predicate logic. We may therefore say that predicate logic is a *concretization* of all possible abstract worlds that satisfy this set of axioms. All axiomatic proofs based on this set of axioms and using the proof rules in Figure 2.1, for example a proof that $\mathsf{T} \vee \mathscr{A} \equiv \mathsf{T}$ or the rules of idempotency, will hold in all concrete worlds that satisfy this set of axioms. In particular, these proofs will hold for predicate logic and for set theory – recall the discussion of the axioms in (2.7) on page 46.

To summarize, we can say that a set of axioms defines a set of abstract worlds – those worlds in which this set of axioms is satisfied. And axiomatic proofs over that set of axioms will infer theorems that hold in *all* such abstract worlds. In Informatics, we may therefore say that axiomatic proofs facilitate proof *reuse*, and this even across seemingly quite different worlds. Remember the example of propositional logic and set theory. The underlying set of five semantic equivalences discussed above are exactly the axioms of so-called *Boolean algebras* (sometimes called *Boolean lattices*). These are important structures in Informatics, and we will discuss them further in the second volume of this trilogy.

Limitations of reasoning in set theory All of modern Mathematics rests on sets of axioms, in particular on axiomatizations of set theory. The latter has different axiomatizations that, in and of themselves, are worthy of mathematical investigation. Indeed, such studies can sometimes reveal principled limitations of axiomatic proofs. For example, one can show that a well-defined mathematical problem over an important algebraic structure called *Abelian groups* – the Whitehead problem – is independent of the axiomatic Zermelo–Fraenkel set theory ZFC (which most mathematicians use in their daily work and whose details are not important for following the discussion in this book). This means that in ZFC it is possible neither to prove that the Whitehead problem has a positive answer, nor that the answer is negative: both answers are consistent with the defining set of axioms for ZFC.

The Whitehead problem is phrased in terms of algebraic structures (here Abelian groups) and their constructions (here homomorphisms). The second volume, *Algebraic Thinking*, will introduce and carefully discuss such structures and constructions.

Fortunately, in Informatics we rarely encounter a practical situation in which we cannot find an answer for such principled reasons. However, in Informatics the question often comes up whether we should use axioms that don't seem to have any constructive, algorithmic content. The *Axiom of Choice* is such a debated axiom and is responsible for the "C" in the set theory ZFC. We refer to [47] for further details about this topic.

A set of axioms is most often chosen not because the axioms capture a belief of what is *true* or *false*, but such a set is engineered so that it implies interesting or desired facts. Therefore, such sets may be constructed incrementally: if we cannot prove a desired theorem with the set of axioms, we may have to add some axiom that allows us to prove it. For example, in the axiomatization of set algebra we would like to have the intuitive equations found below the horizontal line in Lemma 2.3; and these equations are indeed derivable from the axioms stated above that horizontal line.

As already indicated, important properties of a set of axioms are

- *Consistency:* the set of axioms cannot derive a sentence as well as its logical negation, and

- *Completeness:* all true facts expressible in the language of the set of axioms are derivable from these axioms.

Our discussion of the Whitehead problem already suggests that there are limits to securing these properties. In Section 2.3.4, we will discuss further that sets of axioms of sufficient complexity will neither allow us to prove their consistency nor to realize their completeness.

We record that, even though Mathematics is the foundation of many of the natural sciences, it is in and of itself not bound to reflect a correct or accurate image of reality. In fact, many mathematicians as well as the famous late informatician Edsger Dijkstra[19] seem to think that aesthetics plays a more important role than pragmatics.

One may now wonder whether Informatics has a similar stance regarding reality, aesthetics, and pragmatics. In this context, it is useful to point out that what is seen as a *practical* problem and what is considered to be a *real* solution (be it a model, a simulation, an algorithm, etc.) lies in the eye of the beholder. This is something that Informatics shares with Mathematics, where it is often debatable whether some Mathematics is pure or applied, and where the answer to this may change over time.

Like Mathematics, Informatics has the same opportunity to create new realities which are in part self-supporting (for example computer games) or which influence our real world in very significant ways (for example Facebook, Google, Wikipedia, and so forth).

2.3.3 Algebraic Thinking

Sets of axioms introduce a higher level of abstraction than that of a particular problem domain; for example, we saw that the same set of axioms may apply to different logics or structures. Axiomatic proofs therefore operate at a meta-level and apply to

---

[19] See http://en.wikipedia.org/wiki/Edsger_W._Dijkstra

all concretizations of the meta-level, i.e., to all logics or structures that satisfy the set of axioms.

The laws of propositional logic and set theory given in Lemmas 2.1 and 2.3 can be proved with the same set of axioms and so these laws pertain not just to these two worlds but to the aforementioned Boolean algebras. The latter structures are completely characterized by the laws found above the horizontal lines of these two lemmas. Therefore, other laws of Boolean algebras can be derived from these laws.

Boolean algebras play a key role in Informatics. At the technical level, Informatics strongly benefits from binary representations of Boolean algebras and the efficient implementation of their operations. In the second volume of this trilogy, we will gain a deeper understanding of Boolean algebras and, in doing so, will develop typical and useful algebraic methods such as axiomatic proofs, abstraction through maps that preserve structure (so-called *homomorphisms*), and systematic ways of constructing structures – such as products, closures, homomorphic image, and sub-structures.

2.3.4 Soundness and Completeness

In the second volume of this trilogy, we will rigorously exploit the opportunity to organize proofs in a *compositional* manner. This will make use of the structure of an expression (for example, a formula) as well as of the principle of extensionality. The latter allows us to study sub-expressions in isolation. The reason that this approach works so well is that the considered sets of axioms, regardless of whether their axioms are based on rules or equations, satisfy two important meta-properties:

- *Soundness:* We can *only* derive valid statements from the application of inference rules (for example of the proof rules in Figure 2.1) to the set of axioms.
- *Completeness:* *All* valid statements can be obtained from the set of axioms by application of the inference rules (for example by the substitution of equals for equals).

These properties of soundness and completeness are by no means a given for systems of axiomatic reasoning. Let us consider the case of predicate logic over the naturals, which turned out to be unexpectedly problematic, a fact that, when revealed by Kurt Gödel, led David Hilbert to resign from his active research.

Gödel's Incompleteness Theorems The soundness of inferences for predicate logic can be shown for all usual systems of inference. However, the completeness of predicate logic depends strongly on the structures under consideration. For the structure of natural numbers, Gödel was able to show his famous *First Incompleteness Theorem*[20] saying that there cannot be a system of inference based on finitely many rules that is both sound and complete for predicate logic over the natural

---

[20] See http://en.wikipedia.org/wiki/Gödel's_incompleteness_theorems

numbers. The scope of this theorem extends correspondingly to axiomatic reasoning for all sets of axioms that subsume within them a faithful description of the natural numbers. Another consequence thereof is that for the sets of axioms used in Mathematics, for example the set theory of Zermelo–Fraenkel (ZF or ZFC), there are statements that can neither be shown to be true nor shown to be false; see our discussion of the Whitehead problem on page 56.

Gödel's incompleteness theorems had a profound impact on Logic and Mathematics. In 1900, the famous mathematician David Hilbert (* 1862; †1943) created a list of 23 important open problems in Mathematics with the aim of encouraging the research community to direct their attention to these problems. The second problem on that list, called Hilbert's program[21], asked for a proof of the consistency of Mathematics and later developed into the desire to find a finite set of axioms from which all true facts of arithmetic could be derived. Gödel's Incompleteness Theorems proved the impossibility of such a program and were in exact opposition to the intuition that formed the basis of Hilbert's program. With the arrival of these Incompleteness Theorems, Hilbert retired from active research.

On the positive side, many of the algebraic structures that are important in Informatics – let us mention here lattices, groups, rings, fields, and vector spaces – have sound and complete axiomatizations and we will explore these structures further in this trilogy. However, Informatics is concerned not just with soundness and completeness of reasoning, but also with *decidability*: we want automatic (algorithmic) means of deriving true statements or of proving that a given statement is true. Mathematics often considers settings where decidability is unattainable; even in Informatics many practical problems (for example those found in the analysis of programs [50]) are also undecidable. However, in Informatics, we often trade off precision for decidability: in program analysis, for example, we may ask whether a certain piece of code is "dead", i.e., can never be reached in any execution. In general, like many other program analyses, this is undecidable, but most state-of-the-art compiler systems incorporate powerful corresponding optimizations.

It can be shown that answering this simple question is undecidable in general: there is no algorithm that can detect *all* such dead code for *all* programs, and this can be proved with mathematical rigor. However, this does not say that we cannot invent algorithms that identify *some* dead code for *some* programs and where the quality of such algorithms may be measured by how much dead code they can detect for programs of interest. Abstraction turns out to be a key technique in deriving such algorithms: it allows us to trade off the precision of actual program behavior with the ability to efficiently identify dead code in programs. We refer to texts on *program analysis* [55] for further information on this important topic.

---

[21] See http://en.wikipedia.org/wiki/Hilbert's_program

2.3.5 Antinomies

The set theory formulated by Cantor encounters logical limits when applied naively. Bertrand Russell, for example, formulated what is now known as *Russell's Antinomy*: he asked what happens if we define a "set" of all sets that do not contain themselves as element. Formally, we may want to define this set as follows:

$$R =_{df} \{S \mid S \notin S\}$$

Russell then asked whether R itself is a member of R or not! He noted that, according to the definition of R, we have that

$$R \in R \ \ if \ and \ only \ if \ \ R \notin R$$

which is apparently a logical contradiction.

Russell's Antinomy crucially exploits the ability of naive set theory to make self-references in constructing sets such as R above. We saw the problematic aspect of such self-references already in the formulation of paradoxical "propositions" as seen in Example 2.2(3). We had successfully ruled out such statements as proper propositions by insisting that propositions must allow the *consistent* assignment of truth values. In set theory, one can take a similar but more formal approach. For example, the *axiomatic set theory* ZFC provides a formal framework for preventing constructions such as the one for set R above. Concretely, this is done by introducing *types* into the axiomatic language: there are now two types, *sets* and *classes*. Consistent constructions then map sets into sets, but constructions that may lead to inconsistencies will form classes. This creates a *two-level language* for set theory and this separation of levels realizes the desired consistency. This approach of using types to prevent "bad" things from happening is also at the heart of typed programming languages, where types often provide run-time guarantees of the values of expressions but may also prescribe more sophisticated program behavior.

Another antinomy was proposed by Cantor himself, documenting that he was well aware of this problematic aspect of naive set theory: consider the "set" of *all* "sets" $\mathfrak{A}$. Since every subset of $\mathfrak{A}$ also has to be an element of $\mathfrak{A}$, it then follows that the power set $\mathfrak{P}(\mathfrak{A})$ is a subset of $\mathfrak{P}(\mathfrak{A})$. This is a contradiction since Cantor proved that the power set of a set S (be it finite or infinite) is always *strictly* larger (in Mathematics: has larger *cardinality*) than the set S itself. We will explore this proof and its relations to Informatics further in Chapter 3 (Theorem 3.6 on page 92).

Russell's Antinomy has a direct relationship to Informatics. A fundamental question in Theoretical Informatics is to understand which problems can be – in principle – solved through the use of programmable machines. It turns out that not all problems can be solved in this manner, one example being the *Halting Problem*. This problem asks whether we can write a computer program H that takes as input the source code of a computer program P and an input I to program P and decides whether P, when run on input I, terminates or not. It can be shown that there is no program H that can do this; we thus say that the Halting Problem is *undecidable*. The undecidability of this problem leads to the undecidability of many other important problems by indirect proofs: we assume that problem Q is decidable and can then use this to prove that the Halting Problem is decidable; therefore problem Q must be undecidable as the Halting Problem is undecidable; this is a nice example of the use of the Proof Principle of Proof by Contradiction. Let us point out that there is an interesting pedagogical way of relating Russell's Antinomy and the Halting Problem.[22]

2.4 Learning Outcomes

The study of this chapter will give you the learning outcomes discussed below, and the ability the answer the questions below correctly and with confidence.

- A solid appreciation of propositional and predicate logic:

  - What are propositions?
  - How can we construct formulas of propositional or predicate logic?
  - How may we show the logical validity of a proposition?
  - How can we transform propositions into other propositions without changing their semantic meaning, for example through axiomatic proofs?

- A good comprehension of the basic concepts of set theory:

  - What are sets?
  - How can we define sets?
  - In which relationships can sets be to each other?
  - Which operations are defined over sets?
  - What limitations are there on the concept of a set?

- You will have recognized the significance of patterns and structures occurring in Mathematics:

  - What is the relationship between propositional logic and predicate logic?
  - In the context of the previous question, what advantage does the consideration of underlying patterns have over the consideration of concrete logical and set-theoretic operators?
  - What are axiomatic systems?
  - Which requirements are of interest for such axiomatic systems?

- You will also gain a first feeling for the following:

  - That the Compositionality Principle and the Principle of Substituting Equals for Equals are significant and powerful even in simple contexts already.
  - The principle difference between argumentations conducted purely at the syntactic level and argumentations done solely within a semantic level, and what advantages proof methods offer that are syntax-oriented and axiomatic.

---

[22] See http://www.scs.ryerson.ca/~mth110/Handouts/PD/russell.pdf

2.5 Exercises

Exercise 2.1 (Size of truth tables).
Show that a formula of propositional logic that contains $k > 0$ atomic propositions has a truth table with 2^k rows.

Exercise 2.2 (Modelling with propositions).
The lecture *Mathematics for Informaticians* finishes with a written exam which is offered on two different dates, an early and a late date. No student is allowed to take the exam on both dates. Alex, Bea, and Chris want to take this exam, and so need to decide on which date to take it. Here are their constraints:

(a) If Chris takes the exam on the early date, then Alex will have the courage to also take the exam on the early date (otherwise, Alex will choose the late date).

(b) Bea and Chris do job-sharing on a job that helps them with financing their studies. Therefore, at most one of them can take the exam on the early date.

(c) Alex and Chris have made a bet with their common-room mates that at least one of Alex and Chris will take the exam on the early date.

(d) Bea wishes to get a ride in Alex's car when traveling to the exam. Therefore, she has decided that she will take the exam on the early date if and only if Alex takes the exam on the early date.

1. Model the propositions (a) - (d) as expressions of propositional logic. In doing so, make use of propositional atoms $\mathscr{A}, \mathscr{B}, \mathscr{C}$ (for Alex, Bea, and Chris, respectively). These propositions have truth value tt if the corresponding person takes the exam on the early date; otherwise that truth value is ff.
2. Examine the truth tables of these four propositions and determine an assignment that makes all four propositions true.
3. State which of Alex, Bea, and Chris should take the exam on the early date in order to meet all the above four constraints.

Exercise 2.3 (Proving semantic equivalences).

1. Prove the semantic equivalences stated in Lemma 2.1.
2. Prove the semantic equivalence in the second item of Lemma 2.2.

Exercise 2.4 (Non-example of semantic equivalence).
Consider the formulas $\exists x. \forall y. (x = y) \vee (x < y)$ and $\forall y. (0 = y) \vee (0 < y)$ of predicate logic.

1. Show that these formulas have the same truth value over the structure of natural numbers $\mathbb{N}$, and justify this truth value.

2. Show that these formulas are, however, not semantically equivalent. Hint: find a structure with an interpretation of 0, of +, and of <, in which one of these formulas is true, and the other one is false.

Exercise 2.5 (Compositionality and satisfiability).
Recall that a formula $\mathscr{A}$ of propositional logic is *satisfiable* if there is an assignment that makes $\mathscr{A}$ true. Show the following:

1. $\mathscr{A} \vee \neg \mathscr{A}$ is satisfiable for all formulas $\mathscr{A}$ of propositional logic.
2. There is no formula $\mathscr{A}$ of propositional logic for which $\mathscr{A} \wedge \neg \mathscr{A}$ is satisfiable.
3. Show that if one of the formulas $\mathscr{A}$ or $\mathscr{B}$ is satisfiable, then $\mathscr{A} \vee \mathscr{B}$ is satisfiable.
4. Show that similar results do not hold for $\neg$ and $\wedge$: find formulas $\mathscr{A}$ and $\mathscr{B}$ of propositional logic that are both satisfiable such that neither $\neg \mathscr{A}$ nor $\mathscr{A} \wedge \mathscr{B}$ are satisfiable.

Exercise 2.6 (Axiomatic proofs).

Let $\mathscr{A}, \mathscr{B}, \mathscr{C}$ be propositions. Prove the semantic equivalence:

$$\neg\Big(\mathscr{A} \wedge \big(\mathscr{B} \vee \neg(\mathscr{C} \vee \neg \mathscr{A})\big)\Big) \;\equiv\; \mathscr{A} \Rightarrow (\neg \mathscr{B} \wedge \mathscr{C}) \tag{2.10}$$

by using the laws of propositional logic from Lemma 2.1. In doing so, you should think of the implication not as a genuine logical operator but as an abbreviation:

$$\mathscr{A} \Rightarrow \mathscr{B} =_{df} \neg \mathscr{A} \vee \mathscr{B} \tag{2.11}$$

– justified by the fact that (2.11) is a semantic equivalence.
Hint: Transform the expression on the left-hand side of $\equiv$ in (2.10) so that negations only occur in front of atomic propositions; make use of De Morgan's Laws and the Law of Double Negation to achieve this. Then continue to transform the resulting formula, using further laws of propositional logic. Document and thus justify each such application of laws in those transformations.

Exercise 2.7 (Semantic equivalence for exclusive-or).
Recall that the proposition $\mathscr{A} \oplus \mathscr{B}$ is true when exactly one of $\mathscr{A}$ and $\mathscr{B}$ is true. Show the following semantic equivalences:

1. $(\mathscr{A} \oplus \mathscr{B}) \oplus \mathscr{C} \equiv \mathscr{A} \oplus (\mathscr{B} \oplus \mathscr{C})$.
2. $(\mathscr{A} \oplus \mathscr{B}) \oplus \mathscr{B} \equiv \mathscr{A}$.

Exercise 2.8 (Interpretation of axioms in sets and propositional logic).

1. We interpret the axioms in (2.7) in set theory so that variables denote sets, $\leq$ means subset inclusion $\subseteq$, and $\sqcap$ means the intersection of sets $\cap$.

   a. Show that the axioms in (2.7) hold under this interpretation.
   b. Demonstrate that you can prove the equation $x \sqcap x = x$ from the axioms in (2.7), and argue why this also proves that $A \cap A = A$ for all sets A.

2. We interpret the axioms in (2.7) in propositional logic: $\leq$ is interpreted as implication $\Rightarrow$, and $\sqcap$ as logical conjunction $\wedge$.

   a. Show that the axioms in (2.7) hold under this interpretation.
   b. Argue why the last item also implies the semantic equivalence $\mathscr{A} \wedge \mathscr{A} \equiv \mathscr{A}$.

Exercise 2.9 (Empty intersection of family of sets).
Find an example of a family of sets $\mathfrak{S}$ such that $\bigcap \mathfrak{S}$ equals $\emptyset$ but where for all A and B in $\mathfrak{S}$ we have $A \cap B \neq \emptyset$.

Exercise 2.10 (Semantic proofs).

1. Give a semantic proof of the set-theoretic law

$$A \backslash (B \cup C) = (A \backslash B) \cap (A \backslash C)$$

   by appealing to Definition 2.7 for the meaning of these set operators and by using the Equational Laws of Set Operations of Lemma 2.3.
2. Recall that the proposition $\mathscr{A} \oplus \mathscr{B}$ is true when exactly one of $\mathscr{A}$ and $\mathscr{B}$ is true. Prove the semantic equivalence $\mathscr{A} \oplus \mathscr{B} \equiv \neg(\mathscr{A} \Leftrightarrow \mathscr{B})$ by using the Propositional Laws of Lemma 2.1. In doing so, you should think of the logical equivalence not as a genuine logical operator but as an abbreviation:

$$\mathscr{A} \Leftrightarrow \mathscr{B} =_{df} (\mathscr{A} \Rightarrow \mathscr{B}) \wedge (\mathscr{B} \Rightarrow \mathscr{A})$$
$$\mathscr{A} \Rightarrow \mathscr{B} =_{df} \neg \mathscr{A} \vee \mathscr{B}$$

Exercise 2.11 (Symmetric difference and exclusive-or).
Show that for all sets A and B, their symmetric difference $A \triangle B$ equals the set

$$\{x \mid (x \in A) \oplus (x \in B)\}$$

where $\oplus$ denotes the logical operator exclusive-or.

Exercise 2.12 (Goldbach's Conjecture in Predicate Logic).
Express Goldbach's Conjecture (recalling Example 2.1(4) on page 21) as a formula of predicate logic. In doing so, you may introduce suitable auxiliary predicates, similarly to how we used the auxiliary predicate *gcd*.

Chapter 3
Relations and Functions

> *The step that leads from numbers to functions is like the step*
> *that leads from games to strategies*

The mathematical constructions of thought become more and more interesting as the complexity of the relationships between defined objects increases. In the previous chapter, we got to know the divisibility relationship, which expresses that one natural number is a divisor of another natural number. We also introduced set-theoretic operations that express how new sets can be formed from existing ones. The focus of this chapter is on relations and functions. In a manner of speaking, we will therefore formalize and generalize the conceptual relationship expressed in the divisibility relationship.

It will be beneficial to present two concepts of relationships, relations and functions on the one hand, and operations on relations and functions on the other hand. We will illustrate these concepts right away by means of examples. The second volume, *Algebraic Thinking*, will then deliver a deeper and more systematic introduction to operations and their use.

3.1 Relations

Relations, as mathematical objects, express relationships between the elements of sets. Relations are defined in terms of the foundational concept of *Cartesian Product*, which we now discuss.

3.1.1 Cartesian Product

Intuitively, the *Cartesian Product* of two sets A and B is the set obtained by combining each element of set A with each element of set B.

Definition 3.1 (Cartesian product). Let A and B be sets. The *Cartesian Product* of A and B is defined as.

$$A \times B =_{df} \{(a,b) \mid a \in A \wedge b \in B\} \qquad\qquad \square$$

Elements of the form (a,b) in $A \times B$ are called *pairs* or (more precisely) may also be referred to as *ordered pairs*, since the order in which these elements appear matters. In particular, $A \times B$ is generally not the same set as $B \times A$. We can define equality of ordered pairs through equality of their components. Formally, we say that

$$(a,b) = (a',b') \Leftrightarrow_{df} a = a' \wedge b = b'$$

In general, we therefore have that $(a,b) \neq (b,a)$. This inequality distinguishes ordered pairs (a,b) from the corresponding sets of two elements $\{a,b\}$. Recall that for sets the order of their elements plays no role – for example, we have that $\{1,2\} = \{2,1\}$ whereas $(1,2) \neq (2,1)$.

Example 3.1 (Cartesian Product). Consider the sets $A = \{\clubsuit, \spadesuit, \heartsuit, \diamondsuit\}$ and $B = \{Ace, King, Queen, Jack, 10, 9, 8, 7\}$. The Cartesian product $A \times B$ then represents a stripped deck of 32 cards, a French-suited deck of cards for card games such as *Piquet* or the German game *Skat*:

$$
\begin{aligned}
A \times B = \{ & (\clubsuit, Ace), \ldots, (\clubsuit, 7), \\
& (\spadesuit, Ace), \ldots, (\spadesuit, 7), \\
& (\heartsuit, Ace), \ldots, (\heartsuit, 7), \\
& (\diamondsuit, Ace), \ldots, (\diamondsuit, 7) \}
\end{aligned}
$$

In case that set A equals set B in Definition 3.1, we often write A^2 instead of $A \times A$. We can also generalize the Cartesian product to more than two sets as follows[1]:

$$M_1 \times M_2 \times \ldots \times M_n =_{df} ((\ldots (M_1 \times M_2) \times \ldots) \times M_n)$$

In the second volume, *Algebraic Thinking*, we will see that the Cartesian product is associative. This means that we may omit some parentheses when writing such expressions. For example, we may then simply write $(m_1, m_2, \ldots, m_n)$ instead of $((..(m_1, m_2), \ldots), m_n)$, and we may speak of *tuples* of length n, or more briefly of *n*-tuples, when referring to elements of such sets. Analogous to the notation A^2, we may also write the shorter A^n instead of $\underbrace{A \times \ldots \times A}_{n \text{ times}}$.

Cartesian Products will reveal themselves in the second volume – *Algebraic Thinking* – as an important construction principle of algebraic structures, especially for vector spaces.

Side note: It is possible to represent ordered pairs solely by means of set-theoretic operations discussed in the previous chapter:

---

[1] In Chapter 4 we will see that the structure of parentheses seen here results from the formal, inductive definitional structure of general Cartesian Products.

$$(a,b) =_{df} \{\{a\},\{a,b\}\}$$

The above nesting of set formations allows us to unambiguously reconstruct the order of these elements, which thus faithfully represents an ordered pair. For example, we may represent the ordered pair $(1,2)$ by $(1,2) = \{\{1\},\{1,2\}\} \neq \{\{2\},\{1,2\}\} = (2,1)$. This representational idea can generally be extended to n-*tuples*, i.e., to constructs of the form

$$(a_1,a_2,\ldots,a_n)$$

that order n elements as discussed.

One consequence of this is that we do not increase the expressiveness of set theory by adding an explicit construction of ordered pairs of n-tuples. Everything we can express through such ordered pairs or tuples we may also express in pure set theory that does not have such pairs or tuples as primitives.

  We may also put this differently: the introduction of n-tuples allows us to use a simpler and more usable representation of a concept through concepts that already exist in set theory. In particular, we can say that the underlying *Theory* of which propositions are true or false (under suitable interpretations) is not changed by the introduction of n-tuples.

The purpose of the notation for n-tuples is thus not to express a genuinely richer concept but to build a useful concept out of more primitive concepts in set theory. To put this in the language of Informatics, we view a *tuple language* as a *domain-specific extension* of the language of set theory which allows the treatment of problems in which tuples play a prominent role, for example bitvector procedures that support more efficient implementations of set-theoretic operations through characteristic functions; and this extension is expressible within the language of set theory itself.

Finding representations that are adequate for specific problem domains is a central concern in Informatics. The adequacy of such representations depends not only on *what* we wish to represent, nor on *what* operations we wish to execute over them. But we also want to specify and control *who* may execute these operations – for example in order to control the access to sensitive data files. The separation of syntax and semantics discussed in Section 4.3.2 plays a central role in such specifications and control mechanisms. For example, many security problems in computer systems arise from a confusion of data and programs. Both are pieces of syntax but they have different semantics or intent.

> **Excursion: Finite bitvectors** In Informatics, a very important example of the n-fold Cartesian Product is that over the set of bits $\{0,1\}^n$. The elements of $\{0,1\}^n$ are also called *bitvectors* of length n. Further reflections on this structure can be found in Section 3.4.

3.1.2 n-ary Relations

An n-ary relation may be defined as follows.

Definition 3.2 (n-ary relation). Let $n \geq 1$ be given as well as sets $M_1,\ldots,M_n$. A subset R of set $M_1 \times \ldots \times M_n$ is called an *n-ary relation* on set $M_1 \times \ldots \times M_n$. $\qquad \square$

As a convention, n-ary relations are also called ternary relations when n equals 3, binary relations when n equals 2, and unary relations when n equals 1. Relations play a significant role in many aspects of informatics. In Figure 3.1, we see an illustration of a binary relation, which associates employees of an organization with their organizational roles. Note that an employee may have more than one organizational role. For example, Brigitte's job contains a management as well as an administrative component. Moreover, binary or n-ary relations may not associate elements at all. For example, the organizational role CEO, Chief Executive Officer, is not associated with any of these individuals. This may mean that the CEO is some other individual not featuring in the first set; but it may also mean that this organization does not have such a role or does not presently have it assigned to an employee.

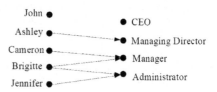

Fig. 3.1 A binary relation that associates employees with their organizational roles.

Let us now explore what unary relations, i.e., 1-ary relations are. By Definition 3.2, a unary relation R is nothing but a subset of M_1. That is to say, unary relations on a set M are just subsets of M. Of course, unary relations are useful. For example, think about the question of whether or not a given natural number is prime, or the question of which people suffer from type 2 diabetes.

Relations are intimately connected to the predicates of predicate logic introduced already in Section 2.1.2. For example, the divisibility relation $n|m$ defined on page 32 is indeed a binary relation on the set of natural numbers. Similarly, the predicate $gcd(n,m,x)$ expresses that x is the greatest common divisor of n and m, and so specifies a 3-ary relation on the set of natural numbers.

3.1.3 Binary Relations

Subsequently, we will essentially focus our attention on *binary* relations. We typically write $R \subseteq A \times B$ to denote such binary relations on the set $A \times B$. In that case, we call set A the *domain* of relation R, whereas set B is called the *co-domain* of relation R. If we invert the roles of domain and co-domain but retain the relation itself, we arrive at the notion of an *inverse relation*.

Definition 3.3 (Inverse relation). Let $R \subseteq A \times B$ be a binary relation. The inverse relation $R^{-1} \subseteq B \times A$ of R is a relation on $B \times A$ and defined by $R^{-1} =_{df} \{(b,a) \mid (a,b) \in R\}$. □

We may compose binary relations provided that the co-domain of one of them equals the domain of the other one. In that case, we may define composition as follows.[2]

Definition 3.4 (Composition of binary relations). Let $R_1 \subseteq A \times B$ and $R_2 \subseteq B \times C$ be binary relations. The *relational composition* of R_1 and R_2, denoted by $R_2; R_1 \subseteq A \times C$, is defined by:

$$R_2; R_1 =_{df} \{(a,c) \mid \exists b \in B. \ (a,b) \in R_1 \wedge (b,c) \in R_2\} \qquad \square$$

The construction of relational composition is illustrated in Figure 3.2. From Definition 3.4 and that figure we can see that this composition turns relations on sets $A \times B$ and $B \times C$ into a relation on set $A \times C$. In that composition, set B has "disappeared": its role in $R_2; R_1$ is merely to link elements of A to elements of C via connecting links of set B.

The so-called *identity relation* on set A, denoted by I_A, has special significance for relational composition.[3] It is defined as:

$$I_A =_{df} \{(a,a) \mid a \in A\}$$

For any binary relation $R \subseteq A \times B$, we then have $R; I_A = R = I_B; R$. That is to say, the relational composition of a binary relation with the corresponding identity relation won't change the underlying binary relation.

Convention: Instead of writing $(a,b) \in R$, we will often make use of the so-called *infix notation* $a R b$, which has the same meaning as $(a,b) \in R$. This convention is often applied for binary relations using symbols such as "$=$","$\leq$","$<$", and so forth, and you most likely have encountered and used infix notation such as $3 = 2 + 1$ in school already. Occasionally, you may see expressions such as $(2,3) \in \ \neq$ which of course has the same meaning as $2 \neq 3$.

We now define the concepts of image and co-image of a binary relation.

Definition 3.5 (Image and co-image of a binary relation). Let $R \subseteq A \times B$ be a binary relation. For an element a in A, its *image under relation R* is defined as

$$R(a) =_{df} \{b \in B \mid (a,b) \in R\}$$

Analogously, for an element b in B we define its *co-image under relation R* as

$$R^{-1}(b) =_{df} \{a \in A \mid (a,b) \in R\}$$

$$\qquad \square$$

---

[2] We use semicolon as the operator symbol for relation composition as this emphasizes the evaluation from left to right. This is somehow similar to programming, where the semicolon usually means the sequential composition of program parts. Historically, function composition (cf. Equation (3.1)) is defined in the opposite direction, i.e., the function that is being applied first is written to the right of the composition operator ∘.

[3] If A is clear from context, we may write I instead of I_A subsequently.

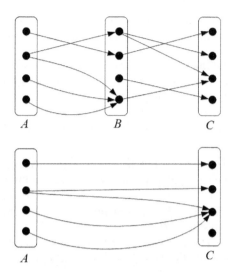

Fig. 3.2 Relational composition

The notation for image and co-image suggests that the image of a under relation R is the same as the co-image of a under the inverse relation R^{-1}. We leave it as an exercise to prove this fact.

In the definition above, it should be noted that both the image and the co-image may be an empty set. For example, for R from Figure 3.1 we see that $R(John)$ as well as $R^{-1}(CEO)$ are empty. The notation $R(a)$, respectively $R^{-1}(b)$, is usually reserved in Mathematics for the case when R is a function, a concept that we will explore next. However, its use for relations can result in more compact formulations of propositions. For example, the two formulas $\forall b \in R(a). \mathscr{A}(b)$ and $\forall (x,y) \in R. x = a \Rightarrow \mathscr{A}(y)$ of predicate logic are equivalent.[4] Finally, in software engineering one often writes $R(a)$ in the form $R.a$ and then speaks of *relational navigation* of a via relation R.

A special role is played by binary relations R over sets $A \times A$, i.e., by binary relations whose domain and co-domain are equal. We call those binary relations *homogeneous*. You already know examples of homogeneous relations, the identity relations I_A. On closer inspection, this is the restriction of the equality relation $=$ to set A. So I_A and I_B are different relations if A and B are different even though they both only relate equal elements. In programming languages, this phenomenon is called *polymorphism*. For example, the homogeneous relation $\leq \subseteq \mathbb{R} \times \mathbb{R}$ is defined over the real numbers but gives rise to many restricted versions of $\leq$ on sets $A \times B$ where A and B are subsets of $\mathbb{R}$. Another example of a homogeneous relation is the complement of the equality relation, "$\neq$".

---

[4] This is another illustration of the choice of suitable domain-specific notation or representation.

3.2 Functions

In school mathematics, *functions* are introduced as an unambiguous association of elements of the domain with elements of the co-domain. The use of the terms "domain" and "co-domain" suggests that functions are special kinds of relations, and this is indeed the case. Intuitively, a function associates each element of the domain with a unique element of the co-domain. For example, the social security numbers of all citizens should ideally represent a function: all citizens should have such a number (for their pensions and so forth), and no citizen wants to have the same such number as another citizen (which may lead to erroneous pension payments, for example). We may formalize this intuitive concept of functions through binary relations that enjoy special properties. Before doing so below, it is useful to consider two symmetrically conceived properties of uniqueness and totality, which we define next.

Definition 3.6 (Left and right uniqueness).   A binary relation $R \subseteq A \times B$ is

1. *right unique* iff

$$\forall a \in A, b_1, b_2 \in B. \ (a, b_1) \in R \land (a, b_2) \in R \ \Rightarrow \ (b_1 = b_2)$$

2. *left unique* iff

$$\forall a_1, a_2 \in A, b \in B. \ (a_1, b) \in R \land (a_2, b) \in R \ \Rightarrow \ (a_1 = a_2) \qquad \square$$

Right unique relations are those for which no element of the domain is related to different elements of the co-domain. For example, the relation *has-genetic-mother* is right unique as everyone can have at most one genetic mother. Figure 3.3 shows a relation that is *not* right unique (the arrows shown in boldface violate this property). Similarly, left unique relations are those for which elements of the co-domain are not related to different element of the domain. The relation *has-bank-account-number* is expected to have this property in the real world: surely, we would not want the same bank account number to be related to two different people. Figure 3.4 shows a relation that is right unique, but not left unique. Indeed, the inverse relation *bank-account-number-belongs-to* may be another such example as people typically have more than one bank account with different account numbers.

A relation for which all elements of the domain are related to one or more elements of the co-domain is called left total. Dually, if all elements of the co-domain are related to one or more elements of the domain, the relation is called right total. An intuitive example of a left total relation is *has-father* over the set of human beings (assuming for a moment that this set does not change over time). This is so since every human being must have a father "by construction" – although developments in genetic engineering may change that in the future. An intuitive example of a right total relation is the relation *father-of*, which is just the inverse of relation *has-father*. This suggests that inverting a relation changes left totality into right totality, and vice versa.

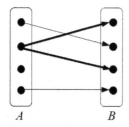   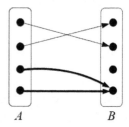

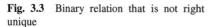

A   B   A   B

Fig. 3.3 Binary relation that is not right unique

Fig. 3.4 Binary relation that is not left unique

Definition 3.7 (Left and right totality). A binary relation $R \subseteq A \times B$ is

1. *left total* $\Leftrightarrow_{df} \forall a \in A. \exists b \in B. (a,b) \in R$
2. *right total* $\Leftrightarrow_{df} \forall b \in B. \exists a \in A. (a,b) \in R$

$\square$

The relation in Figure 3.3 is evidently right total but not left total, whereas the relation in Figure 3.4 is left total but not right total.

Based on some of the properties of relations introduced above, we can now formalize the familiar concept of a function.

Definition 3.8 (Function). A right unique and left total relation is called a *function* (or sometimes also a *map*). $\square$

For a relation $f \subseteq A \times B$ that is indeed a function in the sense just defined, one often uses the notation $f : A \rightarrow B$ instead. Consider an element a of the domain of f. Since f is left total, we know that there is some b in the domain of f such that $(a,b) \in f$. But since f is also right unique, we infer that there can be only one such b. Therefore, we are entitled to write $b = f(a)$ instead of $(a,b) \in f$ subsequently.

However, note that for an arbitrary element b^* of the co-domain of f there may be no, one, or more than one element a of the domain such that $f(a) = b^*$. For example, consider the function $f(x) = \max(0,x)$ as a function of the real numbers, mapping from $\mathbb{R}$ to $\mathbb{R}$. For -1 in the co-domain there is no x in $\mathbb{R}$ such that $f(x) = -1$. For 1, there is exactly one element x in $\mathbb{R}$ (namely $x = 1$) such that $f(x) = 1$. And for 0, there are infinitely many x (namely all non-positive real numbers) such that $f(x) = 0$. This is not surprising since the inverse of a function f, f^{-1}, is generally only a relation and not a function itself. The association of function values with elements of the domain is often also written in the form $a \mapsto f(a)$.

Given sets A and B, we can now form the set of all functions $f : A \rightarrow B$, and we write B^A to denote this set. For example, let A be $\{1,2,\dots,n\}$. Then B^A has the same number of elements as the Cartesian product B^n. To see this, note that every n-tuple $(b_1,\dots,b_n) \in B^n$ can be thought of as a function with domain $\{1,\dots,n\}$ and co-domain B given by the mapping $i \mapsto b_i$. The notation B^A therefore generalizes this from sets of the form $\{1, \ldots, n\}$ to arbitrary *index sets*, even to infinite ones. It is a fact of set theory that there are $|B|^{|A|}$ functions from set A to set B. In other words, we have that $|B^A|$ equals $|B|^{|A|}$.

Analogous to the product relation, introduced in Definition 3.4, we may define the *composition* of functions $f: A \rightarrow B$ and $g: B \rightarrow C$ as the chaining or successive application of two functions. It should be noted that left totality and right uniqueness are preserved by such a composition, making this therefore well defined for functions:

$$g \circ f =_{df} f; g \tag{3.1}$$

where the ; on the right-hand side is the relational composition as defined in Definition 3.4 of the two functions interpreted as binary relations between their input and output. We can understand this notation when we apply it to an element a of A, the domain of function f:

$$(g \circ f)(a) = g(f(a))$$

The identity relation I_A is, by construction, also a function from A to A, i.e., A is domain and co-domain. We already saw that it acts as an identity for relational products. Therefore since identity relations are also functions, identity relations are also identities for functional composition. Subsequently, we write id_A for I_A when we want to emphasize that this is a function. For functions $f: A \rightarrow B$ we then have that $f \circ \mathrm{id}_A = \mathrm{id}_B \circ f = f$. Function id_A is also called the *identity function* (or simply the *identity*) on A.

Whenever we have $A \subseteq B$, this determines a unique *inclusion function* $i_{A,B}: A \rightarrow B$ via the map $i_{A,B}(a) =_{df} a$ for all a in A.[5] It is useful to formalize the notion of function restriction.

Definition 3.9 (Function restriction). Let $f: A \rightarrow B$ be a function and $A' \subseteq A$ and $B' \subseteq B$ be subsets of A and B, respectively.

1. The restriction of f to domain A' is denoted by $f|_{A'}: A' \rightarrow B$ and given by $f|_{A'}(x) = f(x)$ for all x in A'.
2. Dually, suppose that $f(x)$ is in B' for all x in A. Then the restriction of f to co-domain B' is well defined, denoted by $f|^{B'}: A \rightarrow B'$, and given by $f|^{B'}(x) = f(x)$ for all x in A. □

Below, we will also write $f|^{B'}_{A'}$ to mean the function that first restricts the domain to A' and then the co-domain to B'; of course, the latter is only allowed if $f(x)$ is in B' for all x in A'. Note that the function restriction $f|_X$ could equivalently be defined as $f \circ i_{X,A}$.

---

[5] Informaticians would perhaps speak here of an "(up-)cast" from set A to set B, for example when casting an integer value to a real value.

3.2.1 Properties of Functions

Left totality and right uniqueness are the defining properties of functions, whereas their symmetric counterparts, right totality and left uniqueness, are important properties that functions may enjoy. Studying this leads to the concepts of injectivity, surjectivity, and bijectivity of functions. Intuitively, injectivity means that no two elements are mapped to the same element; surjectivity means that the function reaches all elements of the co-domain; and bijectivity means that the inverse of the function is again a function on the co-domain of the inverted function.

Definition 3.10 (Injectivity, surjectivity, and bijectivity). A function $f \colon A \to B$ is:

1. *injective* $\Leftrightarrow_{df}$ f is left unique;
2. *surjective* $\Leftrightarrow_{df}$ f is right total;
3. *bijective* $\Leftrightarrow_{df}$ f is injective and surjective.   □

Let us consider some examples of these properties.

Example 3.2.

1. The function $f_1 \colon \mathbb{N} \to \mathbb{N}$ with $n \mapsto 2n$ is injective, since for $n, m \in \mathbb{N}$ with $n \neq m$ it follows that $f_1(n) \neq f_1(m)$. On the other hand, f_1 is not surjective, since there is no $n \in \mathbb{N}$ with $f_1(n) = 1$.

2. The function $f_2 \colon \mathbb{Z} \to \mathbb{N}$ with $z \mapsto |z|$ is surjective,[6] since every $n \in \mathbb{N}$ is the image of a corresponding number from $\mathbb{Z}$. However, f_2 is not injective, since $f_2(-1) = f_2(1) = 1$. This means that the number 1 is "hit" by two different elements of $\mathbb{Z}$.

3. The function $f_3 \colon \mathbb{Q} \to \mathbb{Q}$ with $q \mapsto 2q$ is bijective. Its injectivity follows analogously to that for f_1. Its surjectivity results from the fact that every rational number $q \in \mathbb{Q}$ is of the form $\frac{q'}{2}$ for some rational number q' in $\mathbb{Q}$.

Our argument above that f_1 is injective did not directly appeal to the definition of left uniqueness, which we may express through the implication

$$f_1(a_1) = f_1(a_2) \implies a_1 = a_2$$

Rather, the above proof used the implication

$$a_1 \neq a_2 \implies f_1(a_1) \neq f_1(a_2)$$

This approach is generally known as the ***Proof Principle of Contraposition***:

---

[6] $|z|$ denotes the absolute value of z, defined by $|z| =_{df} \begin{cases} z & \text{if } z \geq 0 \\ -z & \text{otherwise} \end{cases}$. We make use of the notation $|\cdot|$ within this trilogy by overloading, as usual, this symbol to refer to the absolute value, the cardinality of sets, or the length of words in a formal language.

Proof Principle 4 (Contraposition)

Let $\mathscr{A}$ and $\mathscr{B}$ be propositions.

$$(\mathscr{A} \Rightarrow \mathscr{B}) \equiv (\neg\mathscr{B} \Rightarrow \neg\mathscr{A})$$

In other words, we may prove the implication $\mathscr{A} \Rightarrow \mathscr{B}$ by proving the contraposed implication with negated propositions.

The correctness of the Proof Principle of Contraposition may be shown by inspecting the truth tables of these two implications, which we leave as an exercise.

The injectivity and surjectivity of functions is preserved under functional composition. Specifically, we have the following.

Theorem 3.1 (Invariance Theorem). *Let* $f : A \rightarrow B$ *and* $g : B \rightarrow C$ *be functions. Then we have:*

1. $g \circ f$ *is injective, whenever* f *and* g *are injective.*

2. $g \circ f$ *is surjective, whenever* f *and* g *are surjective.*

3. $g \circ f$ *is bijective, whenever* f *and* g *are bijective.*

Proof. Item (3) of the theorem follows directly from items (1) and (2).

For item (1), consider two elements $x, y \in A$ with $(g \circ f)(x) = (g \circ f)(y)$. According to the definition of functional composition, this means that $g(f(x)) = g(f(y))$. Because of the injectivity of g, this first implies that $f(x) = f(y)$. Now, we appeal to the injectivity of f and infer from $f(x) = f(y)$ that x equals y.

For item (2), let c be an arbitrary element of C. Since g is surjective, there is some element b in B with $g(b) = c$. Since f is surjective, there is also an element a in A such that $f(a) = b$. But for this element a we then obtain $(g \circ f)(a) = g(f(a)) = g(b) = c$ as desired. □

We already dealt with quantified propositions and their proofs. But we would like to recall here, and make explicit, the principal approach to proving quantified propositions. To that end, we consider the above proof of the surjectivity of $g \circ f$. Formally, this is proved by showing the truth of the following quantified proposition – which nests a universal with an existential quantifier:

$$\forall c \in C. \, \exists a \in A. \, (g \circ f)(a) = c$$

The core of the proof of Theorem 3.1 consists in the following **Proof Principle of the Elimination of Quantifiers**:

Proof Principle 5 (Elimination of Universal Quantifiers)

The universal quantifier of a universally quantified proposition $\forall x \in M.\ \mathscr{A}(x)$ is eliminated by replacing variable x with an element of M that is assumed to be **arbitrary but fixed**, *and then proving the proposition $\mathscr{A}(x)$ for that element of M.*

In the proof of $\mathscr{A}(x)$, we may only assume about x what we could assume for an arbitrary but fixed element of M, i.e., what properties we could derive for x from the definition of set M. If M is the natural numbers, for example, we may only assume that x is a number that is integral and non-negative. For those M that are defined inductively, for example, as the set of lists of integers say, we often have to do a case analysis for such an element (a list of integers) that appeals to the inductive definition of set M. We will explore the latter further in Chapter 6.

In such a proof, we often use formulations such as

- *"Let $x \in M$ be arbitrarily chosen (but fixed)"*

or in a short form we may just say

- *"Let x be in M."*

Proof Principle 6 (Elimination of Existential Quantifiers)

The elimination of an existentially quantified proposition $\exists y \in M.\ \mathscr{A}(y)$ can be achieved by deliberately choosing for y a **targeted and specific** *element of M, and then showing the proposition $\mathscr{A}(x)$ for this specific element of M. The choice of this element is typically tailored to the needs of proposition $\mathscr{A}$.*

In the proof of proposition $\mathscr{A}(x)$ we may then use all properties enjoyed by the targeted and specific element of M. For example, let us consider the proof of $\exists y \in \mathbb{N}.\ (y \leq 4) \wedge \exists x \in \mathbb{N}.\ y = x^2$, which says that there is a natural number y less than or equal to 4 that is also a square of a natural number. We can eliminate the outermost existential quantifier by choosing y to be 4. Now we have to prove the proposition $(y \leq 4) \wedge \exists x \in \mathbb{N}.\ y = x^2$ with y being 4, Since $4 \leq 4$ is true, it remains to prove that $\exists x \in \mathbb{N}.\ y = x^2$ is true when y is 4. Here, we eliminate x by setting it to 2. Finally, we have to show that proposition $y = x^2$ is true with y being 4 and x being 2 – which is clearly the case. Note that, although we eliminated these two quantifiers in sequence, the choice 4 made for y already anticipated the choice 2 we would make for x. Such a "look-ahead" is not uncommon in the elimination of nested quantifiers.

In a proof that uses existential elimination, we typically use formulations such as

- *"Let y be given as ..."*

or

- *"Set $y =_{df} ..."*

We discuss the usage of existential quantifier elimination in examples subsequently.

We can see an application of the Proof Principles 5 and 6 for quantifier elimination in the proof of the surjectivity of $g \circ f$ above in the two formulations "let c be an arbitrary element of $C\ldots$" and "But for this element a we then obtain$\ldots$" The latter formulation makes reference to an implicit choice of element a such that $f(a) = b$. That there is such an element a is a consequence of the surjectivity of f. In a way, the choice of a is made by appealing to some property (surjectivity of f) that gives us a desired a. In contrast, the former formulation does indeed choose an arbitrary but fixed element c – which is required since this eliminated a universal quantification.

We note that the assumptions of injectivity, respectively surjectivity, for f and g in items (1) and (2) of Theorem 3.1 are sufficient for proving the injectivity, respectively surjectivity, of $g \circ f$. However, these assumptions are not necessary for proving the respective property of $g \circ f$. For example, suppose that the domain A of f contains only one element. Then $g \circ f$ is injective regardless of whether or not g is injective. A finer analysis reveals that the proof for item (1) still goes through for injective f, provided that function g is injective on a restriction of its domain to the image of f – the set $\{f(a) \mid a \in A\}$. Similarly, the proof of item (2) is still valid if there are subsets $X \subseteq A$ and $Y \subseteq B$ such that both $f|_X^Y : X \to Y$ and $g|_Y : Y \to C$ are surjective. Note that $f|_X^Y$ restricts f in its domain via X, and in its co-domain via Y. In particular, for this to be well defined we need that $f(x)$ is in Y for all x in X.

The following combinatorial result leads to other insights regarding the properties of injectivity, surjectivity, and bijectivity – and can basically be derived by a combination of common sense and counting with our fingers.

Theorem 3.2. *Let A and B be finite sets with $|A| = |B|$, and $f : A \to B$ a function. Then the following three statements are equivalent:*

1) f is injective, 2) f is surjective, and 3) f is bijective.

These results are quite intuitive and accessible, not least because they are rooted in the so-called *Pigeonhole Principle*, which says that it is impossible to put n pigeons into m pigeonholes when $m < n$. This domain assumes that no two pigeons fit into a sole pigeonhole. Of course, this principle applies to any sets of objects, not just pigeons. This principle is also called Dirichlet's Drawer Principle, named after the mathematician who proposed it in the nineteenth century. We can use this principle, for example, to derive that at a concert attended by 380 fans there are at least two fans at that event who share their birthday: a year does not have more than 366 days and there are more than 366 fans at the event.

A dual principle is the one that says that some drawers must remain empty if there are fewer objects to put into drawers than drawers themselves. For example, if only 350 fans attend the concert, then there is at least one day of the year that is not the birthday of any attending fan. We can describe both of these principles with mathematical precision as follows:

Proof Principle 7 (Pigeonhole Principle)

Let A and B be finite sets and $f : A \to B$ a function. Then we have:

1. *Whenever $|A| > |B|$, then f is not injective.*

2. *Whenever $|A| < |B|$, then f is not surjective.*

The proof of this Pigeonhole Principle is easy, but requires as a tool a **Proof Principle of Induction**, which we will cover in Chapter 5, where we will return to proving the Pigeonhole Principle in Exercise 5.22).

Below, we will make use of standard notation for a function $f: A \to B$ which we now introduce for subsets $X \subseteq A$ and $Y \subseteq B$:

$$f(X) = \{f(x) \mid x \in X\} \qquad f^{-1}(Y) = \{a \in A \mid f(x) \in Y\} \qquad (3.2)$$

In words, $f(X)$ is the image of the restriction $f|_X$, whereas $f^{-1}(Y)$ is the *pre-image* of Y under f – those elements of A that f maps into set Y. It should be clear that $f^{-1}(B) = A$, $f^{-1}(\emptyset) = \emptyset$, and that $f(A)$ equals B iff f is surjective.

Proof. (of Theorem 3.2) It is apparent that it suffices to show that the injectivity of f implies the surjectivity of f, and vice versa.[7] To that end, we prove each of these two implications as follows:

"$\Rightarrow$":   We show this implication via Proof Principle 4 of Contraposition. Assume that f is not surjective. We have to show that f is also not injective. Since f is not surjective, there is some element b in B with $b \notin f(A)$. Let $B' =_{df} B \backslash \{b\}$. Then we have $|B'| = |B| - 1 = |A| - 1 < |A|$. Because of the Pigeonhole Principle 7.1 we infer that $f|^{B'} : A \to B'$ is therefore not injective. But this implies that f with co-domain B is also not injective.

"$\Leftarrow$":   This implication is also proved via the Proof Principle of Contraposition. Let f be not injective. We need to show that f is also not surjective. Since f is not injective, there exist two different elements a_1, a_2 in A with $f(a_1) = f(a_2)$. We consider now the restriction of f to the domain $A' = A \backslash \{a_1\}$, that is the function $f|_{A'} : A' \to B$. Then we have that $|A'| = |A| - 1 = |B| - 1 < |B|$. Appealing to the Pigeonhole Principle 7.2 this implies that $f|_{A'}$ is not surjective. But since we also have that $f(a_1) = f(a_2)$, we conclude that f with domain A itself is also not surjective. $\square$

Theorem 3.2 is of course also true in cases in which the domain and co-domain are given by the same finite set A. The bijections $\pi : A \to A$ rearrange all elements of

---

[7] Usage of the word "apparent" or similar expressions such as "as is easily seen" or "without loss of generality" should be encountered with caution if not suspicion, as they typically hide some details of a complete proof and these omitted details may well be incorrect. In this case, though, we are justified to say that these implications suffice: if we prove that f if injective iff it is surjective, then this means that it is also injective iff it is bijective – since the latter is essentially the conjunction of the other two properties.

A in a unique manner, and any such unique rearrangement is a bijection on A. One often speaks then of *permutations* π. The study of permutations is important in Informatics. For example, in the optimization of algorithms we often have *symmetric* situations where these symmetries are witnessed by permutations; and we can then often compute a result for just one of these situations as a representative result for all those situations. Permutations are also of interest in Mathematics; in the second volume of this trilogy, we will see that they form the structure of a *group* through the composition of functions.

3.2.2 Invariants

The establishment of Archimedean Points is a fundamental concept in the analysis of a new or modified structures, situations, or applications [73]. As soon as we begin with the introduction of operators, functions, transformations, or constructors, the question of Archimedean Points becomes central. For example, what should be the Archimedean Point of a compiler, a program that translates source code (another program) into machine code (yet another program)? We claim that this should be the *semantics*, in the sense that the functionality of a program is fully preserved when translated from source code to machine code.[8] A fundamental concern is also when Archimedean Points are maliciously or unintentionally corrupted. For example, a compiler may include optimizations in its machine code generation that boost performance but that may not preserve semantics. This may cause a lot of problems, such as wrong results, program abortion, and perhaps even worse, difficult to detect security vulnerabilities of the running machine code.

Similar considerations can be made when encrypting messages. The Archimedean Point would be that we can, in principle, recover the original message from the encrypted one. Another Archimedean Point here may be that it should be impossible for someone to recover the original message from the encrypted one without knowing the used encryption key; this "invariant" is assumed, for example, in the qualitative analysis of cryptographic protocols but it may have to be weakened to a stochastic invariant that replaces *impossibility* with *small probability*.

We now explore suitable Archimedean Points for two operations on functions that we discussed in this chapter already: functional composition, and the inverse relation of a function (seen as a relation). The Invariance Theorem 3.1 showed that functional composition not only preserves the property of being a function, but also the more specific function properties of injectivity, surjectivity, and bijectivity. On the other hand, we already saw that the inverse relation f^{-1} of a function f is generally not a function. For example, $f \colon \mathbb{N} \to \mathbb{N}$ given by $f(x) = (x \bmod 2)$ maps natural numbers to their parity (which captures whether they are even or odd), whereas f^{-1} relates 0 to all even numbers, 1 to all odd numbers, and other numbers to no

---

[8] What we mean by the semantics of programs and how one can prove the preservation of such semantics is the subject of topics in Informatics such as Compiler Construction, Semantics of Programming Languages, and Formal Methods for System Design.

numbers. Another example of this is function f_2 of Example 3.2. This begs the
question of why functions are not generally preserved by the inverse relation, and
answering this question should lead to the discovery of specific conditions under
which the inverse of a function is also a function. Such an analysis will not only create a better understanding of these relationships, but it will typically lead to concrete
proposals for better design.

Consider a function $f: A \to B$ that is injective, i.e., it is right unique. Then we
know that the inverse relation f^{-1} is left unique. But since f is not necessarily
surjective, there may be an element in B for which f^{-1}, seen as a function, is not
defined. This is illustrated in Figure 3.5. From this figure we can intuitively see that
the restriction of f^{-1} to domain $f(A)$ is a function since f is injective. In particular,
if $f(A)$ equals B (meaning if f is surjective), f^{-1} is that restriction and so also a
function.

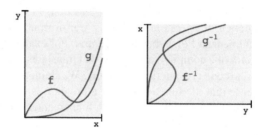

Fig. 3.5 On the left is the graph of a non-injective function $f: A \to B$ as well as the graph of an
injective function $g: A \to B$. On the right are the graphs of the corresponding inverse relations f^{-1}
and g^{-1}, respectively. While f^{-1} is not even a function, the injectivity of g makes the restriction
$g^{-1}|_{g(A)}: g(A) \to A$ a function. However, $g^{-1}: B \to A$ is a function only if g is surjective, i.e., if
$g(A) = B$

Our concept of a function is sometimes called a *total* function, since $f: A \to B$
is defined for all elements of A. It is sometimes useful to consider a variant of that
concept, so-called *partial* functions $f: A \hookrightarrow B$ such that the domain on which f
is defined is a proper subset of A. This is exactly the situation we found for g^{-1}
in Figure 3.5, which is a partial function with domain of definition $g(A)$. We will
discuss partial functions further in Section 3.2.4.[9]

All bijective functions f have a so-called *inverse function*, which we denote by
f^{-1}. The invertibility of bijective functions singles them out amongst all functions,
and we often use the more compact term *bijection* to refer to a bijective function.
It is less common to use the terms *injection* and *surjection* for functions that are
injective, respectively, surjective.

---

[9] Partial functions are not really much used or liked in Mathematics. In Informatics, however,
partial functions are a routine and accepted tool. We find them in the semantic description of
programming languages, and in the description of algorithms that may not always terminate, to
name two prominent examples.

Theorem 3.3 below expresses an interesting uniqueness property of inverses of bijections. The proof of this theorem, asked for in Exercise 3.22, illustrates the usage of Proof Principles 5 and 6 for the elimination of quantifiers.

Theorem 3.3. *Let $f : A \to B$ and $g : B \to A$ be functions with $g \circ f = \mathrm{id}_A$ and $f \circ g = \mathrm{id}_B$. Then both f and g are bijective. Furthermore, we have that $f^{-1} = g$ and $g^{-1} = f$.*

In the second volume of this trilogy, *automorphisms* will play an important role; automorphisms are bijective functions that also preserve the given algebraic structure. In fact, one of the most beautiful theories of Mathematics – Galois Theory – is based on the study of so-called automorphism groups, which are certain algebraic structures in which the elements are automorphisms. Galois Theory is not just aesthetically pleasing and elegant. It also has some central consequences, for example the impossibility of angle trisection or of the quadrature of the circle. The appreciation of these consequences would require the introduction of transcendental numbers, which would go far beyond the considerations of this elementary trilogy.

3.2.3 Cardinality of Infinite Sets

In Section 2.2, we introduced the concept of cardinality for finite sets. Essentially, the cardinality is an abstraction of the individual identity of elements in a set. The cardinality of a set is not concerned with *which* elements are found in the set, but solely with *how many* elements are in the set. For finite sets, the determination of their cardinality therefore amounted to little else but a careful count of all their elements. It is perhaps a surprising fact, and often difficult to comprehend, that this simple counting method will no longer work for the determination of cardinalities of infinite sets.

A popular and effective way of illustrating the strange aspects of cardinalities for infinite sets is that of Hilbert's Paradox of the Grand Hotel[10], also known as Hilbert's Infinite Hotel or just Hilbert's Hotel.[11]

> **Hilbert's Hotel** Imagine a fully booked hotel that features *infinitely many* bookable rooms. We assume, as usual, that rooms are numbered increasingly by room numbers in $\mathbb{N}$, starting with room number 1. Now suppose that an additional guest without a reservation arrives at the hotel reception. The receptionist promises this new guest a room without any hesitation. How does the receptionist have the confidence to make such a offer? Well, she simply asks all current guests of the hotel to move from their room to the one whose

---

[10] See http://en.wikipedia.org/wiki/Hilbert's_paradox_of_the_Grand_Hotel

[11] This paradox also illustrates that the intuitively valid Pigeonhole Principle (see Proof Principle 7) no longer applies to infinite sets.

room number is one larger than their own. Thus, the room with number 1 will become vacant yet all existing guests still have a room.

From this we learn that infinite sets possess properties that cannot be comprehended within our *finite* experience. In this case, we can explain this new effect via an informal invariant: *infinity* = *infinity* + *1*.[12] This invariant expresses that we may add to an infinite set finitely many new elements without changing the *cardinality* of that infinite set.

This insight is certainly useful. Can it guide us to extend the usual theory of finite sets – including its calculations of cardinalities – so that it applies to infinite sets as well yet renders the familiar results for finite sets?

Informaticians are constantly confronted with questions of this type. For example, how may we extend a software system S so that we preserve all its known properties but also fulfill additional properties now sought by clients of that system? Even software companies that dominate their market find it hard to manage their product lines so that they meet such demands: the evolution of a system to a new version will of course bring features that attract new customers and encourage existing ones to stay; but new versions are often produced under time-to-market pressures and often contain flaws that corrupt old features or introduce new security vulnerabilities.

Theoreticians call extensions that fully preserve existing features *conservative*, *seamless*, or *natural* extensions. Although the calculus of cardinalities developed below is a natural extension of finite cardinalities, natural extensions rarely occur "in the wild" such as in areas of software engineering or systems engineering. Newly desired features or properties of planned extensions often stand in conflict with some features or properties of the current system version. This means that we often have to trade off old for new properties, which may undo previous design or implementation decisions and so violate the concept of natural extension. A good example of this is when telecom companies introduced new features such as call forward and call blocking to telephone services. Each feature was written as software for a switchboard and some features interacted in undesired and unpredicted ways. This is therefore also known as the *feature interaction problem*. It is difficult to estimate the economic damage that results from system extensions that break existing features. This damage may be in the form of loss of reputation of the software vendor, direct monetary loss of users (for example when their services are down for some time), indirect monetary loss (for example in terms of the hours of work required to re-install a working version of the system) or even loss of life (for example in the 2015 Seville Airbus A400M Atlas crash).

In Mathematics, we don't seem to have such a chaotic situation since mathematical studies tend not to be the subject of time-to-market pressures. In Mathematics,

---

[12] There is a formal theory for computing with infinities, the so-called ordinal numbers (see `http://en.wikipedia.org/wiki/Ordinal_number`). The addition for ordinal numbers does not satisfy this informal equation, but we also won't have any need for studying ordinal numbers in this trilogy.

everything seems to be constructed elegantly from mathematical building blocks. To be fair, Mathematics had centuries of time to construct such a unique edifice. Informatics is a much younger discipline, and it often finds itself in a situation in which design decisions have to be made prior to knowing exactly what the desired system properties are or will be. This situation has also inspired technically minded graphic cartoonists, as can be seen in a Dilbert comic strip.[13] Therefore, *agility* as the ability to adapt flexibly to new constraints and demands plays a crucial role in Informatics. We can master agility by orienting ourselves on Archimedean Points of a system. Adaptations of systems that are natural extensions are ideal and so preferred, since they extend systems in controlled ways that are void of any *collateral damage*, without compromising existing system behavior.

For the natural extension of the concept of cardinality from finite to infinite sets, it suffices to extend the concept of counting. Instead of associating with each set a natural number as its cardinality, as done in Section 2.2.4 for finite sets, we introduce cardinalities and their order directly, simply via the existence of suitable functions.

Definition 3.11 (Cardinality relationships of sets). Let A and B be sets.

1. A and B have the same cardinality[14], denoted by $A \cong B$, if there is a bijective function $f : A \to B$.

2. A has cardinality *less than or equal to B*, denoted by $A \leq B$, if there is an injective function $f : A \to B$.

3. If the cardinality of A is less than or equal to the cardinality of B but A and B don't have the same cardinality, we say that the cardinality of A is *(strictly) less than* that of B, denoted by $A \lneq B$. $\qquad\qquad\square$

Note that the cardinality of A is strictly less than the cardinality of B if there is an injective function from A to B but there is no such bijective function. By appealing to the Pigeonhole Proof Principle 7, which is valid for finite sets, it should be intuitively clear that the above definition fully and correctly accounts for cardinalities of finite sets. For example, the finite sets $\{1,2,3,4\}$ and $\{\clubsuit, \spadesuit, \heartsuit, \diamondsuit\}$ are equipotent whereas the sets $\{2,3,4,5\}$ and $\{yellow, red, blue\}$ are not. In the latter case, the cardinality of the set $\{yellow, red, blue\}$ is less than the cardinality of the set $\{2,3,4,5\}$.

But what is the effect of the above definition on infinite sets? And how does it relate finite and infinite sets? Well, we can express the principle of Hilbert's Hotel through the bijection

$$f : \mathbb{N} \to \mathbb{N}_{>0}$$
$$n \mapsto n+1$$

between the infinite sets $\mathbb{N}_{>0}$ and $\mathbb{N}$. In particular, $\mathbb{N}_{>0}$ and $\mathbb{N}$ are equipotent even though the former is a strict subset of the latter. But infinite sets are far more flexible

---

[13] See http://dilbert.com/strip/2006-01-29

[14] Sometimes also called *equipotent*, *equipollent*, or *equinumerous*.

when it comes to cardinalities. Before we can demonstrate this, we need a formal definition of infinite sets that makes the principle of Hilbert's Hotel more precise.

Definition 3.12 (Finite and infinite sets). A set M is called *infinite* if and only if

$$\exists M' \subset M.\, M' \cong M$$

Otherwise, the set M is called *finite*. $\square$

In words, this definition says that a set is infinite when it contains a strict subset that is equipotent to the entire set. It turns out that we can prove the equipotency of two sets by proving instead that each one has cardinality less than or equal to the other one. We will get to know this proof principle more generally as the *Proof Principle of Antisymmetry* in the second volume of this trilogy. Fundamental to this is the following *Theorem of Cantor, Bernstein and Schröder*[15].

Theorem 3.4 (Cantor-Bernstein-Schröder). *Let A and B be sets. Then we have:*

$$A \cong B \iff A \lesssim B \land B \lesssim A.$$

Although this fact seems rather intuitive, its formal proof is in no way trivial to accomplish. In the second volume of this trilogy we will present a proof of Theorem 3.4 based on Tarski's fixed-point theorem.

Arguably the most well-known representative of infinite sets is $\mathbb{N}$. The central peculiarity of $\mathbb{N}$ is that all of its elements can be enumerated in an (infinite) linear list – much like the set of rooms in Hilbert's Hotel. This property of an infinite set is often more formally referred to as *countable infinity*. The transition from finite lists that enumerate finite sets to infinite lists seems a minor step to make. But it turns out to exactly characterize the simplest form of infinite sets: a set M is countably infinite if, and only if, it is equipotent to the set $\mathbb{N}$ of natural numbers. This is intuitively evident, since every bijection $f\colon \mathbb{N} \to M$ that realizes $M \cong \mathbb{N}$ can be interpreted as an *enumeration* $f(0), f(1), f(2), \ldots$ of all elements of M. Conversely, from such a bijection f we get its inverse bijection $f^{-1}\colon M \to \mathbb{N}$, which can be interpreted as an *enumeration* $f^{-1}(m_1), f^{-1}(m_2), \ldots$ of all elements of $\mathbb{N}$ through elements of M.

The mathematician Georg Cantor was the first to realize the extent and limitations of countably infinite sets, and he made these realizations tangible via his famous *Diagonalization Argument*. But before we study this argument, let us first consider two variants of Hilbert's Hotel. The first variant considers the combination of two Hilbert Hotels as we have introduced them. This can be achieved by using the set of integers $\mathbb{Z}$, where negative numbers model room numbers of one hotel whereas non-negative numbers (0 or natural numbers) model room numbers of the other hotel.[16]

---

[15] See http://en.wikipedia.org/wiki/Schröder-Bernstein_theorem

[16] Appreciating this approach demands a certain flexibility in interpreting models, which is a core competency of an Informatician, as it is a key for transferring conceptual approaches into concrete applications.

The question then becomes, how can we rebook all guests in both hotels so that they all have rooms in one such Hilbert Hotel – assuming of course that each guest gets his own room? For this, it suffices to find a bijection $h\colon \mathbb{Z} \to \mathbb{N}$ as this will tell us how to rebook each guest: for room number z in $\mathbb{Z}$, rebook the guest in that room to room number $h(z)$ in $\mathbb{N}$ in the new Hilbert Hotel. We can find such an h by first finding its inverse function, which we call $f_{\mathbb{Z}}$ and which is defined as:

$$f_{\mathbb{Z}}\colon \mathbb{N} \to \mathbb{Z}$$

$$n \mapsto \begin{cases} \frac{n}{2} & \text{if } n \text{ is even} \\ -\frac{n+1}{2} & \text{if } n \text{ is odd} \end{cases}$$

The function $f_{\mathbb{Z}}$ maps natural numbers to integers as follows:

$$0 \mapsto 0,\ 1 \mapsto -1,\ 2 \mapsto 1,\ 3 \mapsto -2,\ 4 \mapsto 2, \ldots$$

It is apparent that $f_{\mathbb{Z}}$ is surjective, since every negative integer z equals $f_{\mathbb{Z}}(-(2z+1))$ for the natural number $-(2z+1)$; whereas every non-negative integer z (including 0) equals $f_{\mathbb{Z}}(2z)$ for the natural number $2z$. Moreover, the function $f_{\mathbb{Z}}$ is also injective (and therefore bijective as well): different natural numbers get mapped to different integers – we leave the proof of this claim as an exercise. Therefore, we conclude that $\mathbb{N}$ and $\mathbb{Z}$ are equipotent (and h may be chosen as $f_{\mathbb{Z}}^{-1}$). Mapping this insight into our application domain of Hilbert Hotels, this means that we can rebook all guests of two Hilbert Hotels successfully in a sole Hilbert Hotel without having to put more than one guest into any room.

The above argument generalizes from two to any finite number of Hilbert Hotels (which one could prove formally with the induction techniques developed in Chapter 6). More surprising is that $\mathbb{N} \times \mathbb{N}$ is also equipotent to $\mathbb{N}$. In other words, we may rebook without any double bookings guests of infinitely many Hilbert Hotels in a sole Hilbert Hotel! Figure 3.6 shows the construction of a bijection $d_C\colon \mathbb{N} \times \mathbb{N} \to \mathbb{N}$ that realizes this surprising equipotency. This construction is known as the *Cantor pairing function*. The Cantor pairing function d_C can be given in explicit form:[17]

$$d_C(m,n) =_{df} \frac{1}{2}(n+m)(n+m+1) + m$$

The equipotency of $\mathbb{N}$ and $\mathbb{N} \times \mathbb{N}$ is also of such great importance since we may represent the set of rational numbers $\mathbb{Q}$ as (integral) fractions, i.e., via pairs of integers. Let us make this argument more formal. Since $\mathbb{N}$ and $\mathbb{Z}$ are equipotent, we may restrict our attention to the set of non-negative rational numbers $\mathbb{Q}_{\geq 0}$ and its representation in the countably infinite set $\mathbb{N} \times \mathbb{N}$. Different integral fractions may represent the same rational number. For example, since $\frac{1}{2} = \frac{2}{4} = \frac{3}{6} = \ldots$ there are (infinitely) many functions from $\mathbb{Q}_{\geq 0}$ to $\mathbb{N} \times \mathbb{N}$. Our argument that $\mathbb{Q}_{\geq 0}$ is count-

---

[17] The expression $\frac{1}{2}(n+m)(n+m+1)$ corresponds to $\sum_{i=0}^{n+m-1}(i+1) = \sum_{i=0}^{n+m} i$. The latter is exactly the number of enumerated elements of those diagonals that have been traversed completely, plus the m enumeration steps of the last diagonal.

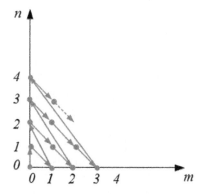

Fig. 3.6 Cantor pairing function: this realizes an enumeration of elements from the set $\mathbb{N} \times \mathbb{N}$ along the marked line through $d_C(0,0) = 0$, $d_C(0,1) = 1$, $d_C(1,0) = 2$, $d_C(0,2) = 3$, $d_C(1,1) = 4 \ldots$

ably infinite won't depend on which of these functions we choose, as long as it is indeed a function and that we can prove that this function is injective. For it to be a function, we need that each element of $\mathbb{Q}_{\geq 0}$ is mapped to a unique element of $\mathbb{N} \times \mathbb{N}$.[18] A canonical choice for such a function d would be to map every element r of $\mathbb{Q}_{\geq 0}$ to the uniquely determined pair (p, q) of relatively prime numbers for which r equals the fraction p/q. The latter we expressed as $gcd(p, q) = 1$ in Chapter 2. We leave it as an exercise to prove that this does indeed specify an injective function $d: \mathbb{Q}_{\geq 0} \to \mathbb{N} \times \mathbb{N}$.

Therefore, we managed to show that $\mathbb{Q}_{\geq 0}$ has cardinality less than or equal to $\mathbb{N} \times \mathbb{N}$, which is equipotent to $\mathbb{N}$. Concretely, the injective function $d_C \circ d: \mathbb{Q}_{\geq 0} \to \mathbb{N}$ realizes that $\mathbb{Q}_{\geq 0}$ has cardinality less than or equal to $\mathbb{N}$.

But this was already the difficult direction of showing that $\mathbb{Q}_{\geq 0}$ and $\mathbb{N}$ are equipotent when following the proof pattern suggested in Theorem 3.4. The other direction, that the cardinality of $\mathbb{N}$ is less than or equal to that of $\mathbb{Q}_{\geq 0}$, follows directly from the existence of the inclusion function $i_{\mathbb{N}, \mathbb{Q}_{\geq 0}}: \mathbb{N} \to \mathbb{Q}_{\geq 0}$ introduced on page 75.

At this point, it makes good sense to introduce another proof principle – so-called *Circle Inference* – whose general manifestation will be explored further in the second volume of this trilogy.

Proof Principle 8 (Circle Inference – special instance)

Let $n \geq 2$ and $A_1, A_2, \ldots, A_n$ be sets with cardinality relationships $A_1 \lesssim A_2 \lesssim \cdots \lesssim A_{n-1} \lesssim A_n \lesssim A_1$. Then we have that $A_1 \cong A_2 \cong \cdots \cong A_{n-1} \cong A_n$.

---

[18] For this map $r \mapsto (p, q)$ it is immaterial whether or not r equals the fraction p/q. What matters is whether this mapping represents a function that is *injective* and provably so.

According to this proof principle, to show that the three sets $\mathbb{Q}_{\geq 0}$, $\mathbb{N} \times \mathbb{N}$, and $\mathbb{N}$ are equipotent, it suffices to simply show that $\mathbb{Q}_{\geq 0} \leq \mathbb{N} \times \mathbb{N} \leq \mathbb{N} \leq \mathbb{Q}_{\geq 0}$. Compared to the above proof, this approach won't lead to a much simpler argument. On the other hand, we have not yet proven that $\mathbb{Q}_{\geq 0}$ and $\mathbb{N} \times \mathbb{N}$ are equipotent. We could have done that above by another application of Theorem 3.1.1 and Theorem 3.4. In the end, a Circle Inference is nothing but an aid for *proof optimization* so that we can minimize the number of proof obligations and appeals to theorems such as the ones aforementioned.

Remark: The above definition of function d is yet another example of the potential of domain-specific representations. Here we are merely looking for some injective map of type $\mathbb{Q}_{\geq 0} \to \mathbb{N} \times \mathbb{N}$. The derivation of that property was helped by pairing the Principle of *Preservation of Semantics* with the Principle of *Canonical Representation*. The latter made sure that this mapping results in a function and not just a relation. The former made sure that this function is injective.

Let us summarize the cardinality facts we have shown so far:

$$\mathbb{N} \cong \mathbb{Z} \cong \mathbb{N} \times \mathbb{N} \cong \mathbb{Q}$$

In fact, we may extend these insights from $\mathbb{N} \times \mathbb{N}$ to Cartesian products $\mathbb{N}^n$ for an arbitrary but fixed natural number n. Although this can easily be accomplished by means of induction, as developed in Chapter 5, a visualization of the corresponding generalization of the Cantor pairing function to $\mathbb{N}^n$ is already pretty complicated. This illustrates the power of *stepwise and modular approaches*. We first prove a generic result. In this case, we would need to prove that the Cartesian product of two countably infinite sets is countably infinite again.[19] Once this generic result is shown, we may appeal to it iteratively using induction – which we will develop in Chapter 5.

Our discussion suggests that countable infinity is well preserved by (finite) Cartesian products. In fact, one can also show that the countable union of countably infinite sets is countably infinite as well. So are there set-theoretic operations that do not preserve countable infinity? Cantor also asked, and answered, this question by inventing his famous *Diagonalization Argument*: the set-theoretic operation of power sets does of course preserve the property of being a finite set; but it does not preserve countable infinity.[20]

To prove that the power set $\mathfrak{P}(\mathbb{N})$ of set $\mathbb{N}$ is, unlike set $\mathbb{N}$ itself, not countably infinite, it is helpful to appreciate that there is always a bijective function between

---

[19] Such generalizations are part and parcel of any Informatician's daily work so that his or her methods become *reusable* in different contexts or systems.

[20] The power set operation pervades all of Informatics as we often need to represent problems or their solutions by appealing to power sets, for example in constructing so-called deterministic automata in theoretical computer science or in the semantic description of programming languages that support parallel execution. Power sets are expensive as they increase the size of the problem domain exponentially.

the power set $\mathfrak{P}(M)$ of a set M and the set of functions of type $M \to \{0,1\}$. To see this, note that a subset A of M can be represented through its *characteristic function*

$$\chi_A : M \to \{0,1\}$$

$$\chi_A(m) =_{df} \begin{cases} 1 & \text{if } m \in A \\ 0 & \text{otherwise} \end{cases}$$

Conversely, any function $f : M \to \{0,1\}$ gives rise to a subset $f^{-1}(\{1\})$ through its inverse relation. We leave it as an exercise to prove that the function $A \mapsto \chi_A$ has $f \mapsto f^{-1}(\{1\})$ as inverse, and so is indeed bijective. This fact motivates an alternative notation for $\mathfrak{P}(M)$ as 2^M and the latter is often used in Informatics, where we view 2 as the set $\{0,1\}$.

We now have all the tools at our disposal to prove that set $\{0,1\}^{\mathbb{N}}$ (and therefore also the equipotent power set $\mathfrak{P}(\mathbb{N})$ of $\mathbb{N}$) is not countably infinite.

Theorem 3.5. $\mathbb{N} \lneq \{0,1\}^{\mathbb{N}}$.

Proof. Since $\mathbb{N} \lneq \{0,1\}^{\mathbb{N}}$ means $\mathbb{N} \leq \{0,1\}^{\mathbb{N}}$ but $\mathbb{N} \ncong \{0,1\}^{\mathbb{N}}$, we first show that $\mathbb{N} \leq \{0,1\}^{\mathbb{N}}$ is the case. To that end, we define for each n in $\mathbb{N}$ a function $\varphi_n : \mathbb{N} \to \{0,1\}$ as follows:

$$\varphi_n(x) =_{df} \chi_{\{n\}}(x) = \begin{cases} 1 & \text{if } x = n \\ 0 & \text{otherwise} \end{cases}$$

Next, we consider the map $f : \mathbb{N} \to \{0,1\}^{\mathbb{N}}$ defined as

$$f(n) =_{df} \varphi_n$$

It is clear that f is a function. To see that it is injective, we note that $n \neq n'$ implies $\varphi_n \neq \varphi_{n'}$, since we must have $\varphi_n(n) = 1 \neq 0 = \varphi_{n'}(n)$ by the definition of the function family $(\varphi_n)_{n \in \mathbb{N}}$. Therefore, f is an injective function and so $\mathbb{N} \leq \{0,1\}^{\mathbb{N}}$ follows.

The proof of $\mathbb{N} \ncong \{0,1\}^{\mathbb{N}}$ is more complex and rests on the aforementioned *Cantor Diagonalization Argument*. That argument reveals that there cannot be any function from $\mathbb{N}$ to $\{0,1\}^{\mathbb{N}}$ that is surjective. In particular, this then implies that there cannot be a function of that type that is bijective.

This argument relies on Proof Principle 3 of Contradiction: we *assume* that there is a surjective function $g : \mathbb{N} \to \{0,1\}^{\mathbb{N}}$, and then derive a contradiction from this.[21] We may write the type of g equivalently as

$$g : \mathbb{N} \to \underbrace{\{f \mid f : \mathbb{N} \to \{0,1\}\}}_{\{0,1\}^{\mathbb{N}}}$$

---

[21] Note that this also relies on the elimination of the universal quantifier "for all functions f of type …" to consider an arbitrary function of that type, and then to rule out that this can be surjective.

Given that, we may then define a function $h : \mathbb{N} \to \{0,1\}$ as follows:

$$h(n) =_{df} 1 - g(n)(n) \text{ for all } n \in \mathbb{N}$$

Figure 3.7 illustrates the definition of function h where the functions $g(i) : \mathbb{N} \to \{0,1\}$ are represented in rows of the form

$$g(i) = g(i)(0) \; g(i)(1) \; g(i)(2) \ldots$$

Intuitively it is clear that on the diagonal of that representation we have by construction always a point on which the function of the corresponding row differs from the function h. In other words, g cannot be surjective, which contradicts our choice of g. $\qquad\square$

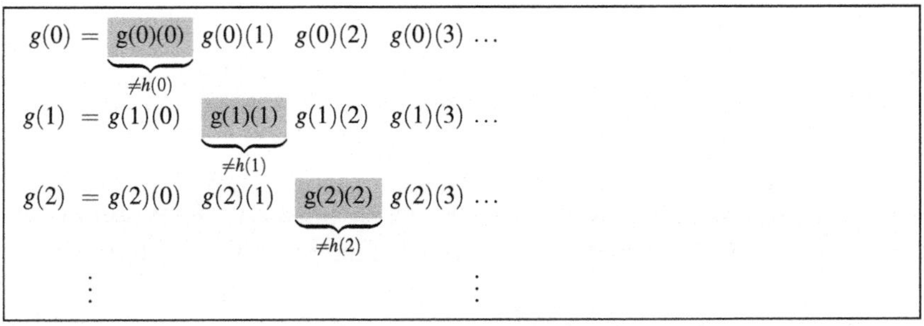

Fig. 3.7 Cantor's diagonalization argument for the non-surjectivity of a function g in the proof of Theorem 3.5. Function h is shown to not be "reached" by g, i.e., to not be in the image of g

We just saw that there are infinite sets that are not countably infinite. There is a standard name for this property.

Definition 3.13 (Uncountable set). Let A be an infinite set. If A is not countably infinite, A is called *uncountable*. $\qquad\square$

You may find this surprising, but it is actually possible to use Cantor's Diagonalization Argument to prove that there is an infinite hierarchy of uncountable sets – meaning that there are uncountable sets $A_0, A_1, \ldots$ such that for all n in $\mathbb{N}$ the cardinality of A_{n+1} is strictly larger than the cardinality of A_n. This, again, is facilitated through a stepwise and modular approach:

- First, we generalize Theorem 3.5 to Theorem 3.6. We leave this as an easy exercise that merely has to systematically adapt the proof steps to that more general setting. Only the case when M equals $\emptyset$ requires its bespoke argument.

- Second, we appeal to Theorem 3.6 repeatedly, and with different instance sets, to keep deriving uncountable sets with strictly increasing cardinalities.

The consideration of such strictly increasing infinite sequences of uncountable sets led to the introduction of *cardinal numbers* in set theory (see Section 3.4.6).

Theorem 3.6. *Let M be an arbitrary set. Then we have that $M \lneqq \{0,1\}^M$.*

The geometric intuitions of most people deteriorate or cease to function in dimension three, which in Mathematics corresponds to the Cartesian product $\mathbb{R}^3$. Physicists and Mathematicians can also think reliably about higher-dimensional spaces such as $\mathbb{R}^{26}$ in String Theory, say. But when it comes to counting, the intuition of most people ceases to function when dealing with countably infinite sets. However, Informaticians should be able to go one step beyond countably infinite sets, by also considering sets of cardinality equal to that of set $\mathfrak{P}(\mathbb{N})$. The latter has pretty interesting equipotent sets, for example we have:

$$\mathfrak{P}(\mathbb{N}) \cong \{0,1\}^{\mathbb{N}} \cong (0,1) \cong \mathbb{R} \tag{3.3}$$

where $(0,1)$ is defined as the set $\{x \in \mathbb{R} \mid 0 < x < 1\}$ and referred to as the open interval of real numbers between 0 and 1.

The leftmost $\cong$ relationship in (3.3) we already showed above. The remaining two equalities of cardinality are proved by Circle Inference (see Proof Principle 8). Choosing a suitable order for traversing the circle we are left to show:

$$\mathbb{R} \leqq (0,1) \leqq \{0,1\}^{\mathbb{N}} \leqq \mathbb{R} \tag{3.4}$$

For the leftmost $\leqq$ relationship in (3.4) we consider the function $f_{\mathbb{R}} : (0,1) \to \mathbb{R}$ defined by

$$x \mapsto \frac{x - \frac{1}{2}}{x(x-1)}$$

We leave it as an exercise to prove that this function $f_{\mathbb{R}}$ is indeed bijective.

The second $\leqq$ relationship, i.e. $(0,1) \leqq \{0,1\}^{\mathbb{N}}$, can be proved based on the observation that every real number strictly between 0 and 1 may be written as a countably infinite bitvector. For example, the fraction $1/3$ equals $0.333\cdots$ in base 10 which can equally be written as a repeating decimal number $0.\overline{3}$ with period 1. Consequently, we represent $0.\overline{3}$ in base 2 as $0.0\overline{10}$ and so as a countably infinite bitvector $010101\ldots$ (or $\overline{01}$ for short). It turns out that this representation is only unique if there is no real number in base 2 with periodic ending $\overline{1}$. For example, in base 2 both $0.010\overline{1}$ and 0.011 represent the same real number. However, to avoid representations with periodic ending $\overline{1}$ it suffices to ensure that the function $f : (0,1) \to \{0,1\}^{\mathbb{N}}$ is injective.

It remains to show that $\{0,1\}^{\mathbb{N}} \leqq \mathbb{R}$. All infinite bitvectors that do not end in the periodic sequence $\overline{1}$ can be mapped to their meaning in $(0,1)$ if interpreted as an infinite sum. For example the bitvector $011\overline{0}$ is mapped to $0.375 = 0 \cdot \frac{1}{2} + 1 \cdot \frac{1}{4} + 1 \cdot \frac{1}{8} + 0 \cdot \frac{1}{16} + \ldots$. All other infinite bitvectors (those with a period ending $\overline{1}$) are mapped to $1 + s$ where s is the value of the infinite sum that the infinite bitvector represents. For instance, the bitvector $010\overline{1}$ is mapped to $1.375 = 1 + 0 \cdot \frac{1}{2} + 1 \cdot \frac{1}{4} + 0 \cdot \frac{1}{8} + 1 \cdot \frac{1}{16} + 1 \cdot \frac{1}{32} \ldots$. This mapping clearly describes an injective function from $\{0,1\}^{\mathbb{N}}$ to $\mathbb{R}$.

3.2.4 Partial Functions

In Mathematics and particularly in Informatics, we often deal with relations that are right unique but not left total. Intuitively, this means that each a in the domain of such a relation f has at most one element b in the co-domain with (a,b) in f: no more than one such b by right uniqueness, and sometimes no such b as left totality does not hold. In Mathematics it is common to write functions such as $inv(x) = 1/x$ from $\mathbb{R}$ to $\mathbb{R}$; this is a function apart from that fact that it is not defined at $x = 0$: it is right unique but not left total. In Informatics, there is yet another source of lack of right totality. Consider a deterministic[22] program that computes the greatest common divisor of two integral inputs. The deterministic nature of the program means that the same inputs yield the same outputs, and so its meaning is a right unique relation. But this relation may not be left total, for example due to a program error in which a while statement gets executed forever and no output is returned.

The functions described above (right unique but not left total) are called *partial functions*. We write $f : A \hookrightarrow B$ to denote that f is a partial function.

For a partial function $f : A \hookrightarrow B$, we denote by $Def(f)$ its *domain of definition*, which is the set $\{a \in A \mid \exists b \in B. \ (a,b) \in f\}$ of all those elements in A that are related to some element in B via f. Note that $Def(f)$ can be any strict subset of A, including the empty set if f is defined nowhere. The restriction of f to its domain of definition converts a partial function into a function. Similarly, we can convert any partial function $f : A \hookrightarrow B$ into a function $f_* : A \rightarrow B_*$ where B_* is $B \cup \{*\}$ with $*$ not in B, and $f_*(a) = f(a)$ if a is in $Def(f)$, and $f_*(a) = *$ otherwise. It is an instructive exercise to show that the set of partial functions $f : A \hookrightarrow B$ is equipotent to the set of functions $g : A \rightarrow B_*$ whenever $*$ is not in B. Note that some authors refer to our functions as *total* functions, since $Def(f)$ is then the total domain of f.

It is of interest to understand how partial functions can be composed with each other. So let $g : A \hookrightarrow B$ and $f : B \hookrightarrow C$ be partial functions. The partial function $f \circ g : A \hookrightarrow C$ has domain of definition $\{a \in Def(g) \mid g(a) \in Def(f)\}$, and for all a in $Def(f \circ g)$ we have $(f \circ g)(a) = f(g(a))$. If we think of total functions $h : C \rightarrow D$ as partial functions with $Def(h) = C$, this also explains how to compose partial with total functions.

---

[22] In such a program, its execution history is uniquely determined by its input values and not, for example, by flipping a coin to determine control flow.

3.3 Equivalence Relations

Equivalence relations capture mathematically the result of reasoning about equality, where the notion of equality used varies from relation to relation and so gives us a means of engineering what we mean by being equal. For example, the relation $\{(r,r) \in \mathbb{R} \times \mathbb{R} \mid r \in \mathbb{R}\}$ is an equivalence relation but it only equates real numbers that are equal as real numbers. Whereas the relation $\{(r,r') \in \mathbb{R} \times \mathbb{R} \mid r-r' \in \mathbb{Z}\}$ is an equivalence relation that identifies two real numbers if their difference is integral, in other words if their meaning which abstracts from the integer part of the real number is equal.

Equivalence relations are therefore seen to be a central tool for the process of abstraction. Other examples of equivalence relations that equate terms of the same meaning are those that relate symbolic expressions such as $x+x$ and $2 \cdot x$ with respect to some theory (here the theory of basic arithmetic), and the relation that relates integer fractions that denote the same rational number, e.g., $\frac{1}{2}, \frac{2}{4}, \frac{3}{6}, \ldots$, to each other.

Formally, equivalence relations are homogeneous relations with the following properties.

Definition 3.14 (Equivalence relation).
   A relation $\sim \subseteq A \times A$ is an *equivalence relation* if, and only if,

1. $\sim$ is *reflexive*, i.e., $\forall a \in A.\ a \sim a$

2. $\sim$ is *symmetric*, i.e., $\forall a_1, a_2 \in A.\ a_1 \sim a_2 \Rightarrow a_2 \sim a_1$

3. $\sim$ is *transitive*, i.e., $\forall a_1, a_2, a_3 \in A.\ a_1 \sim a_2 \wedge a_2 \sim a_3 \Rightarrow a_1 \sim a_3$ ☐

In mathematics, equivalence relations have turned out to be particularly useful for extending numerical domains. In fact, our intuitive reasoning on the fractional representation of rational numbers motivates a formal definition of rational numbers purely based on a relation $\sim_{\mathbb{Q}}$ over $\mathbb{Z} \times \mathbb{Z} \setminus \{0\}$ defined by

$$(a,b) \sim_{\mathbb{Q}} (c,d) \Leftrightarrow_{df} a \cdot d = b \cdot c$$

We leave it as an exercise to prove that $\sim_{\mathbb{Q}}$ satisfies properties (1) – (3) of Definition 3.14.[23]

Based upon this definition one can easily define the addition and multiplication of rational numbers by

$$(a,b) + (c,d) =_{df} (a \cdot d + b \cdot c, b \cdot d)$$
$$(a,b) \cdot (c,d) =_{df} (a \cdot c, b \cdot d)$$

---

[23] It should be noted that $\sim_{\mathbb{Q}}$ would not be an equivalence relation on $\mathbb{Z} \times \mathbb{Z}$, i.e., without restricting denominators to values different from 0. In this case transitivity would be violated, as $(1,0) \sim_{\mathbb{Q}} (0,0)$ and $(0,0) \sim_{\mathbb{Q}} (0,1)$, but $(1,0) \not\sim_{\mathbb{Q}} (0,1)$.

It should be noted that this definition is only reasonable, if it is independent of the particular choice of representatives of rational numbers. For instance, in the case of addition we would have to prove

$$(a,b) \sim_{\mathbb{Q}} (a',b') \wedge (c,d) \sim_{\mathbb{Q}} (c',d') \Rightarrow (a,b) + (c,d) \sim_{\mathbb{Q}} (a',b') + (c',d')$$

Similarly, integers can be constructed upon natural numbers by using an equivalence relation $\sim_{\mathbb{Z}} \subseteq \mathbb{N} \times \mathbb{N}$ which is defined by

$$(a,b) \sim_{\mathbb{Z}} (c,d) \Leftrightarrow_{df} a + d = b + c$$

Here the idea is that non-negative integers are represented by pairs (a,b) with $a \geq b$, whereas negative integers are represented by pairs with $a < b$. For example, the pairs $(5,3)$, $(3,1)$, and $(2,0)$ are all representatives of the positive integer 2, while $(3,5)$, $(1,3)$, and $(0,2)$ refer to the negative integer -2. Like before, the construction is completed by defining a corresponding additional operation:

$$(a,b) + (c,d) =_{df} (a+c, b+d)$$

Example 3.3. We now consider some relations that express relationships between people to see whether these are equivalence relations.

1. The sibling relationship is an equivalence relation if we may assume that everyone is a sibling of himself or herself, and that siblings have the same pair of parents.[24]

2. The friendship relationship is not an equivalence relation, as it is evidently not transitive: Anna may be a friend of Bob, and Bob may be a friend of Charlotte. But Anna may not necessarily be a friend of Charlotte – the stuff of many novels and telenovela.

3. The brother-of relationship is not an equivalence relation, as it is not symmetric: Andrew may be a brother of Beatrix, but Beatrix is a sister and so not a brother of Andrew. However, if we restrict the brother-of relation to the set of all males and assume that every male is a brother of himself (to ensure reflexivity), we obtain an equivalence relation.

An equivalence relation $\sim$ over a set A determines *equivalence classes* on A: for each a in A, we define its corresponding $\sim$-equivalence class $[a]_{\sim}$ as

$$[a]_{\sim} =_{df} \{a' \in A \mid a \sim a'\}$$

In words, $[a]_{\sim}$ is the set of elements of A that are equivalent to a with respect to $\sim$. Since $\sim$ is reflexive, transitive, and symmetric, it is an easy exercise to see that all elements in $[a]_{\sim}$ are equivalent to each other. Moreover, the meaning of the set $[a]_{\sim}$ is independent of the choice of a in that for all a and a' with $a \sim a'$ we have that the equivalence classes $[a]_{\sim}$ and $[a']_{\sim}$ are equal.

---

[24] And so there are neither half sisters nor half brothers.

For equivalence relation $\sim$ over set A, we may form the set of all its equivalence classes. This set is of special interest and will be discussed in more detail next.

3.3.1 Partitions

Partitions are sets of sets that group a set of ground elements into pairwise mutually disjoint sets – where the latter are also referred to as *partition classes*. Formally, we define this as follows.

Definition 3.15 (Partition). Let M be a set. Then a non-empty subset P of $\mathfrak{P}(M)$ is called a *partition of M* $\Leftrightarrow_{df}$

1. All partition classes are non-empty, i.e., $\emptyset \notin P$

2. The partition classes cover set M, i.e.,

$$\bigcup_{M' \in P} M' = M$$

3. The partition classes are pairwise disjoint, i.e.,

$$\forall M_1, M_2 \in P. \ M_1 \neq M_2 \ \Rightarrow \ M_1 \cap M_2 = \emptyset \qquad \qquad \square$$

Let us look at examples and non-examples of partitions.

Example 3.4. For set $M = \{1,2,3\}$, we have that $P_1 =_{df} \{\{1\},\{2\},\{3\}\}$ and $P_2 =_{df} \{\{1,2\},\{3\}\}$ are partitions of M. However, $P_3 =_{df} \{\{1\},\{2\}\}$ is not a partition of M as it does not cover element 3. Similarly, $P_4 =_{df} \{\{1,2\},\{2,3\}\}$ is not a partition of M, since two of its partition classes ($\{1,2\}$ and $\{2,3\}$) are not (pairwise) disjoint.

The intimate relationship between partitions and equivalence relations is captured in the next theorem, expressing that they are merely different representations of the same underlying concept.

Theorem 3.7. *Let A be an arbitrary, non-empty set.*

1. If $P \subseteq \mathfrak{P}(A)$ is a partition on set A, then

$$\sim_P =_{df} \{(a_1, a_2) \in A \times A \mid \exists A' \in P. \ a_1, a_2 \in A'\}$$

is an equivalence relation on A.

2. If $\sim \ \subseteq A \times A$ is an equivalence relation on set A, then the corresponding set of equivalence classes

$$A/\!\sim \ =_{df} \{[a]_\sim \mid a \in A\}$$

forms a partition $P_\sim$ on A.

Proof. 1. In order to prove that $\sim_P$ is an equivalence relation, we have to establish the corresponding three defining properties. Therefore let a_1, a_2, a_3 be arbitrary elements of A and A_1, A_2, A_3 the classes of P that contain a_1, a_2, and a_3, respectively. These classes must exist, because P as a partition covers the entire set A. Then we have $a_i \sim_P a_i$ directly by definition of $\sim_P$, which proves reflexivity. As (different) partition classes are pairwise disjoint, we also know that $a_i \sim_P a_j$ implies $A_i = A_j$, which proves symmetry. Similarly, whenever we have $a_1 \sim_P a_2$ and $a_2 \sim_P a_3$, we can conclude $A_1 = A_2 = A_3$, which proves transitivity.

2. Since a is in $[a]_\sim$ for all a in A, we infer that equivalence classes $[a]_\sim$ are non-empty, and that $\bigcup_{a \in A} [a]_\sim$ equals A and so covers A. It remains to show that equivalence classes are pairwise disjoint.

To that end, consider two elements a_1 and a_2 in A such that their equivalence classes have a non-trivial intersection, i.e., $[a_1]_\sim \cap [a_2]_\sim \neq \emptyset$. Now it remains to show that these equivalence classes are indeed equal, i.e.,

$$[a_1]_\sim = [a_2]_\sim$$

As the intersection is non-empty, there exists some element a' in that intersection, i.e., $a' \in [a_1]_\sim \cap [a_2]_\sim$. Applying the already proved first part of the theorem directly yields as desired $[a_1]_\sim = [a']_\sim = [a_2]_\sim$. $\square$

The above theorem justifies the use of language such as

- an equivalence relation $\sim$ on set A *induced* by partition P on A
- a partition P on set A *induced* by an equivalence relation $\sim$ on A.

In fact, we can prove that for a non-empty set A the mapping $\sim \mapsto P_\sim$ between the set of equivalence relations on set A and the set of partitions on A is bijective and has $P \mapsto \sim_P$ as inverse function:

$$P_{\sim_P} = P \qquad\qquad \sim_{P_\sim} = \sim$$

We leave the proof of this as an exercise. There is also an important relationship between functions and equivalence relations. Every function $f : A \to B$ induces an equivalence relation on set A:

$$a_1 \sim_f a_2 \Leftrightarrow_{df} f(a_1) = f(a_2)$$

The partition corresponding to this equivalence relation is known as the *inverse image partition* of function f; its partition classes are the inverse images of the form $f^{-1}(b)$ where b is in the image $f(A)$ of f.

Example 3.5. Let us consider a set of students

$$S =_{df} \{Adam, Barbie, Chris, Dana, Eric, Fred, Grace, Hannah, Ken, Iris, Jason\}$$

who took part in an exam on Big Data Analytics. We assume that exam grades are from the set $\{A, B, C, D, E, F\}$ where A is the best possible result and F the worst, a "F"ailure. We can model exam grades of these students as a function $g : S \to \{A,B,C,D,E,F\}$, for example:

$$Adam \mapsto B, Barbie \mapsto E, Chris \mapsto A, Dana \mapsto B, Eric \mapsto C, Fred \mapsto C,$$
$$Grace \mapsto C, Hannah \mapsto B, Ken \mapsto F, Iris \mapsto A, Jason \mapsto B$$

The corresponding induced equivalence relation identifies students from set S who have the same exam grade. Here, we obtain the inverse image partition

$$\{\{Chris, Iris\}, \{Adam, Dana, Hannah, Jason\}, \{Eric, Fred, Grace\}, \{Barbie\}, \{Ken\}\}$$

The students who made the top grade are therefore

$$g^{-1}(A) = \{s \in S \mid g(s) = A\} = \{Chris, Iris\}$$

Please note that grade D was not issued for any of these students, and so $g^{-1}(D)$ is empty. Thus this cannot be a partition class; it is not part of that partition as the latter is defined over inverses of elements in the image of g.

The question of whether the intersection or union of equivalence relations is again an equivalence relation often leads to misconceptions.[25] Let us consider the case of intersection first, and let $\sim_1$ and $\sim_2$ be two equivalence relations. In order to show that $\sim_1 \cap \sim_2$, obviously a subset of $A \times A$, is an equivalence relation it suffices to show that it is reflexive, symmetric, and transitive:

- let a be in A; then pair (a,a) is in $\sim_1$ as well as in $\sim_2$ since both relations are reflexive; therefore (a,a) is also in $\sim_1 \cap \sim_2$ and so that relation is reflexive as well;

- let a and b be in A; if (a,b) is in $\sim_1 \cap \sim_2$, then it is also in $\sim_1$ which implies that (b,a) is in $\sim_1$ as well as the latter is symmetric; similarly, we show that (b,a) is in $\sim_2$ and so in the intersection $\sim_1 \cap \sim_2$, which is therefore symmetric.

We leave the proof that $\sim_1 \cap \sim_2$ is transitive as an exercise.

Things present themselves differently, though, when we consider the union of two equivalence relations. We leave it as an exercise to show that the union of two equivalence relations is again reflexive and symmetric (in fact, the proof will not require the transitivity of these relations). What breaks down is the proof that the union is also transitive. So let (a,b) and (b,c) be in $\sim_1 \cup \sim_2$. We would have to show that (a,c) is in that union as well. A case analysis reveals that we cannot prove this when (a,b) is in $\sim_1 \setminus \sim_2$ and (b,c) in $\sim_2 \setminus \sim_1$: it is then possible that (a,c) is neither in $\sim_1$ nor in $\sim_2$.

To illustrate this possibility, let us interpret $\sim_1$ as "having the same (biological) father" whereas $\sim_2$ means "having the same (biological) mother". Let a be Adam, b

---

[25] In this question, we construe equivalence relations on set A as subsets of $A \times A$ and so form their union or intersection.

be Brigitte, and c be Cindy. The above situation means that Adam and Brigitte have the same father (say Ed) but not the same mother, that Brigitte and Cindy have the same mother (say Wendy) but not the same father, and that Adam and Cindy have neither the same father (for example, Cindy's father may be Roger) nor the same mother (for example, Adam's mother may be Margaret).

In this trilogy, we will often see that the intersection is a *benign* operation in that it often preserves structural properties of sets and relations, while union can be problematic.

3.4 Reflections: Exploring Covered Topics More Deeply

3.4.1 From Functions to Strategies

The motto of this chapter was: *The step that leads from numbers to functions is like the step that leads from games to strategies.* We now want to elucidate this by studying the popular and simple game of Tic-tac-toe. Any play in this game starts on an initially empty 3×3 grid of nine cells that will be filled by alternating moves of players X and O, where X makes the initial move in a play. A player makes a move by filling an empty cell of the grid with her symbol X or O, respectively. Any cell can only be filled once in a play. A game is finally won by a player who succeeds in completely occupying a row, column, or diagonal of the grid. A game where all nine cells are filled without reaching a winning situation ends in a draw. Figure 3.8 illustrates some configurations and moves of players in this game.

Let us now formalize the game using sets, relations, and functions. First, we define the set of all *configurations* – the computational states in a play. Formally, Conf is defined as the set of all functions

$$c : \{1,2,3\} \times \{1,2,3\} \to \{\mathsf{X},\mathsf{O},\varepsilon\}$$

The function value $c(i,j)$ captures the entry at row i and column j where ε models an unfilled cell. As a configuration can be graphically understood as a 3×3 matrix filled by elements of the set $\{\mathsf{X},\mathsf{O},\varepsilon\}$, we will write $c_{i,j}$ instead of $c(i,j)$ for convenience.

Next, we identify winning configurations; set Conf_p consists of configurations in Conf that are won by player $p \in \{\mathsf{X},\mathsf{O}\}$:

$$\mathsf{Conf}_p = \{c \in \mathsf{Conf} \mid win(p,c)\}$$

where the predicate $win(p,c)$ expresses the winning condition formally as

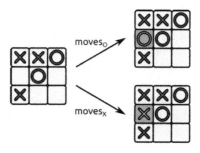

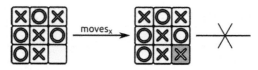

Fig. 3.8 Some configurations and moves of players X and O, indicated respectively by labels $moves_X$ and $moves_O$, between configurations. Note that there are no moves from configurations in which all cells are filled.

$$(c_{1,1} = p \wedge c_{2,2} = p \wedge c_{3,3} = p) \vee$$
$$(c_{3,1} = p \wedge c_{2,2} = p \wedge c_{1,3} = p) \vee$$
$$(\exists i \in \{1,2,3\}. \ c_{i,1} = p \wedge c_{i,2} = p \wedge c_{i,3} = p) \vee$$
$$(\exists j \in \{1,2,3\}. \ c_{1,j} = p \wedge c_{2,j} = p \wedge c_{3,j} = p)$$

We also formalize those configurations where players can still make a move. We call such configurations *enabled*:

$$\mathsf{Conf}_e = \{c \in \mathsf{Conf} \mid \exists i, j \in \{1,2,3\}. \ c_{i,j} = \varepsilon\}$$

A configuration is thus enabled if it contains at least one unfilled cell. The complement $\mathsf{Conf} \setminus \mathsf{Conf}_e$ consists of all configurations in which all cells are filled, and so players can no longer move. We leave it as an exercise to show that no two of the sets Conf_X, Conf_O, and Conf_e are mutually disjoint. Let us denote by $c^0 \in \mathsf{Conf}_e \setminus (\mathsf{Conf}_X \cup \mathsf{Conf}_O)$ the *initial* configuration, where all cells are unfilled, i.e., $c_{i,j} = \varepsilon$ for all i and j. It is important to characterize those configurations from which moves are valid. These are exactly those configurations that are enabled but not won by either of the players yet:

$$\mathsf{Conf}_m = \mathsf{Conf}_e \setminus (\mathsf{Conf}_X \cup \mathsf{Conf}_O)$$

Now, the legitimate moves of player p in $\{X, O\}$ define a relation $moves_p \subseteq \mathsf{Conf}_m \times \mathsf{Conf}$ where

$$(c, c') \in moves_p \Leftrightarrow_{df} next(p, c, c') \tag{3.5}$$

Here the predicate $next(p, c, c')$ is formalized by

$$\exists i, j \in \{1, 2, 3\}. \; c_{i,j} = \varepsilon \; \wedge \; c'_{i,j} = p \; \wedge$$
$$\forall k, l \in \{1, 2, 3\}. \; (k, l) \neq (i, j) \; \Rightarrow \; c_{k,l} = c'_{k,l}$$

Informally, this means that configuration c' differs from c in exactly one cell that has now been occupied by player p. It is worth reflecting that there can be only one combination of i and j in the above existential quantification that makes this predicate true for such a pair (c, c').

A play in this game is now a finite sequence of configurations

$$c^0, \; c^1, \; \ldots, \; c^k$$

such that c^0 is the initial configuration, and for $i < k$ the move (c^i, c^{i+1}) is in $moves_X$ whenever i is even, or in $moves_O$ otherwise. Finally, c^k is the first configuration in this sequence such that $c^k \in Conf \setminus Conf_m$. In other words, players X and O move alternately beginning with player X until a winning situation is reached or all cells are occupied.

We are now prepared to refine the relations $moves_X$ and $moves_O$ to functions that capture *strategies* that these players may *adhere to* in plays. Formally, we may define a *strategy for player* $p \in \{X, O\}$ as a total function

$$f_p : Conf_m \rightarrow Conf \tag{3.6}$$

such that $f_p \subseteq moves_p$ with f_p viewed as a relation.

In a play as in Figure 3.8 player $p \in \{X, O\}$ is said to adhere to his strategy f_X when all moves of p conform with f_p, i.e., $(c_i, c_{i+1}) \in moves_p$ implies $c_{i+1} = f_p(c_i)$.

A strategy f_p is *winning* for player p if all plays in which player p adheres to strategy f_p end in a winning configuration $c^k \in Conf_p$. In other words, a player following a winning strategy must necessarily win, no matter what his opponent plays. It turns out that in the game of Tic-tac-toe there are neither winning strategies for player X nor for player O. However, for both players there are strategies that are *non-losing*. Formally, strategy f_p is non-losing for player p with opponent p' if all plays in which player p adheres to f_p end in a configuration that is *not* in $Conf_{p'}$ – and so is in $Conf_p$ or not in $Conf_e$. To summarize, both players have strategies that are non-losing. And if both players adhere to such strategies, then all plays of the game will end up in a draw, i.e., where c^k is in $Conf \setminus (Conf_e \cup Conf_X \cup Conf_O)$. It is worth noting that this property of the game, the existence of non-losing strategies, will not change if we increase the expressive power of strategies, for example by allowing the determination of the opponent's strategic moves through random coin flips or through the inspection of the history of the current play not just its current configuration. In more general games that relate to formal languages and automata, this is not always the case: random choices or encoding of history in strategies can change the expressive power of such functions [77]. In some games there exist strategies that are winning, but only if they encode the history of the play.

Another interesting observation is that numerous configurations in Conf are not reachable from the initial configuration by legitimate moves. Actually, identifying reachable configurations is a crucial issue in reasoning about the behavior of systems that are based on state transformations. For that reason, reachability analysis is of preeminent importance in model checking [83, 84]. We leave it as an exercise to determine which elements of Conf can be reached from the initial configuration. Furthermore, we leave it as an exercise to determine strategies for both players that are non-losing, for example by writing appropriate programs that explore the relations $moves_X$ and $moves_O$.

3.4.2 Finite Bitvectors

An important example of the n-fold Cartesian product is the set $\{0,1\}^n$. Elements of $\{0,1\}^n$ are often called *bitvectors* of length n. They play a special role in representing subsets of a set $M = \{m_1, \ldots, m_n\}$ of n elements by fixing an enumeration of elements in M (suggested above through the indices m_i). For every subset $A \subseteq M$, its *characteristic bitvector* $b_A = (b_1^A, \ldots, b_n^A)$ in $\{0,1\}^n$ is defined by $b_i^A = 1$ if m_i is in A, and $b_i^A = 0$ otherwise. For example, let M be the set of months in the Gregorian calendar, $M =_{df} \{January, \ldots, December\}$, then we may represent the subset M_{31} of months consisting of 31 calendar days via the characteristic bitvector

$$(1,0,1,0,1,0,1,1,0,1,0,1)$$

which enumerates months in the order in which they appear in the calendar year. You may be familiar with the following *aide-mémoire*, which allows us to reconstruct this bitvector through a rule based on the knuckles of your fingers – as seen in Figure 3.9.

Fig. 3.9 A rule for reconstructing the bitvector that represents the set of 31-day months in the calendar year. Knuckle raises stand for bit value 1, and knuckle slots stand for bit value 0

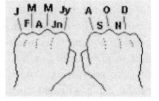

The subset of months M_r that contain the letter "r" in their name[26] can similarly be characterized through the bitvector

$$(1,1,1,1,0,0,0,0,1,1,1,1)$$

---

[26] These months also represent the so-called mussel season in Europe and the east of North America. The fact that M_r captures this season is entirely coincidental. But M_r and its characterizing bitvector represent both of these interpretations.

The representation of sets through bitvectors is of great interest in Informatics, as the relationship between set-theoretic operations and logical operators recorded in Section 2.2.3 suggests. For example, for two subsets M and M' over the same underlying finite set of size n, we can represent the set-theoretic operations of intersection, union, difference, and symmetric difference through the component-wise operations of $b_i \wedge b_i'$, $b_i \vee b_i'$, $b_i \wedge \neg b_i'$, and $b_i \oplus b_i'$ (respectively) on bitvectors. This is particularly efficient since processors may execute such operations for entire machine words (for example, for bitvectors of length $n = 32$ or $n = 64$) with a single instruction. One disadvantage of this representation, at least in its naive realization, is that even a subset of one element or the empty subset require the same space of n bits as the whole underlying set.

3.4.3 Limits of Countable Infinity and Computability

We may use Cantor's pairing function repeatedly to show that the finite Cartesian product of countably infinite sets is countably infinite again. To put this another way, the property of being countably infinite is closed under finite Cartesian products. Let us define the concept of being countable.

Definition 3.16 (Countable set). A set M is *countable* if it is finite or countably infinite.   □

It is not hard to see that the property of being *countable* is also closed under finite Cartesian products. In fact, both properties – countable and countably infinite – have even stronger *robustness*: they are closed under the formation of countable unions of finite Cartesian products.

The proof of countability for finite Cartesian products $\prod_{i=1}^n M_i$ of countable sets M_i uses Cantor's pairing function as before as seen in Figure 3.6 for the case of $n = 2$, where M_1 and M_2 play the role of $\mathbb{N}$ – which is justified as sets M_i are equipotent to $\mathbb{N}$ since they are countably infinite. So how do we extend this proof to the next level of $M_1 \times M_2 \times M_3$? We observe that this set is equipotent to the set $(M_1 \times M_2) \times M_3$ by the associativity of Cartesian products. Cantor's pairing function allowed us to show that $M_1 \times M_2$ is countably infinite as just discussed. So we may think of $(M_1 \times M_2) \times M_3$ as the Cartesian product $M_{12} \times M_3$ where both sets M_{12} and M_3 are countably infinite. Now, we can simply repeat the argument we developed for $M_1 \times M_2$. In the next chapters we will formalize the use of such inductive definitions (for the finite Cartesian products) and inductive proofs (that finite Cartesian products preserve countable infinity).

"Proof tools" such as Cantor's pairing function can often be used in many ways once we have properly understood those tools. For example, we may adjust Cantor's pairing function to prove that the countable union $\bigcup_{i \in \mathbb{N}} S_i$ of countably infinite sets S_i is countably infinite again. Since each S_i is countable, we can write it as an infinite row $a_{i,0}, a_{i,1}, \ldots$ and so get a matrix $(a_{i,j})_{i,j \in \mathbb{N}}$ of infinitely many such infinite rows. The pairing function now interprets each element $a_{i,j}$ as an element (i, j) in

$\mathbb{N} \times \mathbb{N}$ and applies the pairing in this manner to the infinite matrix. This therefore enumerates all elements in $\bigcup_{i \in \mathbb{N}} S_i$ as desired.

We may apply this to the countable union

$$\mathbb{N} \cup \underbrace{\mathbb{N} \times \mathbb{N}}_{\mathbb{N}^2} \cup \underbrace{\mathbb{N} \times \mathbb{N} \times \mathbb{N}}_{\mathbb{N}^3} \cup \ldots \tag{3.7}$$

since all finite Cartesian products of $\mathbb{N}$ with itself are countably infinite as shown above. In Theoretical Informatics, the set in (3.7) is often referred to as the set of finite words over alphabet $\mathbb{N}$. This set is therefore countable for any countably infinite alphabet.[27]

Countability is thus a pretty robust property. However, the countable Cartesian product of countably infinite sets is not countably infinite, and so is *uncountable*: consider the Cartesian product $\prod_{n \in \mathbb{N}} \mathbb{N}$. Its elements are infinite tuples of natural numbers. It is easy to see that $\prod_{n \in \mathbb{N}} \mathbb{N}$ is equipotent to $\mathbb{N}^{\mathbb{N}}$, which according to Theorem 3.5 cannot be countably infinite.

Limits of Computability Let us consider all programs written in a particular programming language, say Java for example. All such programs can be seen as finite words over a finite alphabet, e.g., consisting of the ASCII symbols. Therefore, the set of all Java programs can be written as a countable union of finite sets, similar to the union seen in (3.7), and is therefore countable and infinite (for each n in $\mathbb{N}$ there is a Java program that – as a word – has length at least n). But we saw that there are uncountably many functions from $\mathbb{N}$ to $\mathbb{N}$. Thus, a pure counting argument informs us that there must be functions from $\mathbb{N}$ to $\mathbb{N}$ that cannot be computed by any Java program, or any program in any language over a countable alphabet.[28]

To summarize, we were able to derive from the countability of the set of all programs the existence of functions that cannot be computed by any program. In fact, the set of functions from $\mathbb{N}$ to $\mathbb{N}$ that can be computed by some program is much smaller then the set of such functions that cannot be computed by some program. In mathematical terms, one can measure the sizes of such sets and say that the set of computable functions from $\mathbb{N}$ to $\mathbb{N}$ has measure 0: if you were to randomly pick a function from $\mathbb{N}$ to $\mathbb{N}$, its probability of being computable would be 0. We won't formalize this notion of measure and random choice and merely appeal to their intuitive meaning here.

A special aspect of the argument above is that it is not constructive. We already saw such an example on page 41, when we proved that there are irrational numbers a and b such that a^b is rational. In the above, the non-constructive aspect of the proof is also apparent: even though the vast majority of such functions are non-computable, we do not have a sole example of a non-computable function at hand. We refer to

---

[27] A formal definition of the concept of *word* over a finite alphabet will be given in Section 4.3.1 of the next chapter.

[28] Let us assume for the sake of that argument that the programming language contains a type in which a natural number can be represented, no matter how large it may be.

text books on the Theory of Computability, where concrete and practically relevant such examples are developed by the creation of a rich theory of computability.

3.4.4 Hulls and Closures

We discussed on page 98 that the union of two equivalence relations is not always an equivalence relation. This seems unsatisfactory. For example, consider the equivalence relations $\equiv_3$ and $\equiv_5$ over $\mathbb{Z}$ which relate pairs of integers (x, y) for which the difference $x - y$ is divisible by 3, respectively 5. The union of $\equiv_3$ and $\equiv_5$ is not an equivalence relation: $(5, 2)$ is in $\equiv_3$ and $(2, 7)$ is in $\equiv_5$ but the pair $(5, 7)$ is not in $\equiv_3 \cup \equiv_5$, and so $\equiv_3 \cup \equiv_5$ is not transitive.

This raises an engineering question: can we repair relation $\equiv_3 \cup \equiv_5$ to make it into an equivalence relation, and can we do that with minimal such repairs? This question is of real significance. For example, we may think of $\equiv_3$ and $\equiv_5$ as two different theories for reasoning about equality, and we seek a theory that subsumes both theories and their reasoning but that does not introduce any additional, unnecessary theories in the process. Much of Algebra and Logic is preoccupied with such generalized notions of equality and their combination. We will elaborate and illustrate this kind of reasoning. For example, it is fundamental for mastering the formal semantics of programming languages and in verification of software using tools, such as the Chinese Remainder Theorem in the second book of this trilogy.

Let us now investigate the aspect of repair on the example of $\equiv_3 \cup \equiv_5$. Since both $\equiv_3$ and $\equiv_5$ are reflexive and symmetric, it is easy to see that their union is reflexive and symmetric, as discussed on page 98 already. We could therefore repair this set by adding enough elements to make it transitive, and by ensuring that this addition does not break the other two properties of reflexivity and symmetry. Let us write R for relation $\equiv_3 \cup \equiv_5$ now. Then R^+ denotes the *transitive closure* of relation R. It is uniquely characterized by the following properties:

1. R^+ is a transitive relation over $\mathbb{Z}$
2. R is a subset of R^+
3. for all transitive relations T over $\mathbb{Z}$, we have that $R \subseteq T$ implies $R^+ \subseteq T$.

In other words, R^+ is the smallest transitive relation on the same underlying set that contains R. This pattern for repair can be seen in many algebraic and logical contexts, and it often referred to as *hull*, *closure*, or *completion*. These approaches share the idea of repairing a set by adding the minimal number of elements needed to gain the sought property (in the case of R^+ the property of transitivity). In many applications of this pattern, this repair process uniquely identifies such hulls (for example the so-called convex hull[29] of a set of points in Algebraic Geometry), as is the case for the transitive closure. We leave it as an exercise to show that R^+ equals

---

$$R^+ = \bigcup_{i \in \mathbb{N}_{>0}} R^i \tag{3.8}$$

where R^1 is defined to be R, and for all $i > 1$ it is defined to be $R^{i-1}; R$. This definition follows an inductive pattern similar to the one we will see for defining the Kleene hull on page 137.

We can now prove that $(\equiv_3 \cup \equiv_5)^+$ is indeed the smallest equivalence relation that contains $\equiv_3 \cup \equiv_5$. First, we can prove that $\equiv_3 \cup \equiv_5$ is reflexive and symmetric (as the union of two relations that are reflexive and symmetric). Second, we can reason that the transitive closure R^+ of a reflexive and symmetric relation R is again reflexive and symmetric: this appeals to the definition in (3.8). For reflexivity, we obtain this since R is reflexive and R^1 and therefore R is contained in R^+. For symmetry, we use the symmetry of R to reason that all R^i are symmetric as well; strictly speaking this requires induction, which we study in the next two chapters. But then the union (even an infinite union R^+ in (3.8)) of symmetric relations is symmetric again. To summarize, we now know that $(\equiv_3 \cup \equiv_5)^+$ is an equivalence relation that contains $\equiv_3 \cup \equiv_5$.

To show that $(\equiv_3 \cup \equiv_5)^+$ is the smallest such equivalence relation let T be another equivalence relation over $\mathbb{Z}$ that contains $\equiv_3 \cup \equiv_5$. Exploiting the fact that building transitive closures is monotonic in the sense that the transitive closure of a set always contains the transitive closure of any of its subsets, we get that $(\equiv_3 \cup \equiv_5)^+$ is a subset of T^+ as well. But T is an equivalence relation and so transitive as well. Thus, T^+ equals T and so $(\equiv_3 \cup \equiv_5)^+ \subseteq T$ follows. This proves that $(\equiv_3 \cup \equiv_5)^+$ is indeed the smallest equivalence relation that contains both $\equiv_3$ and $\equiv_5$. There is a chapter in the second volume of this trilogy that elaborates on this kind of reasoning in order to establish the so-called *fixpoint theory*, which can be regarded as the foundation for understanding iterative and recursive programs [56].

In order to strengthen the corresponding intuition, let us now examine the more abstract case, in which we are given a binary relation $R \subseteq A \times A$ over a set A. We would like to know whether there is a unique, minimal equivalence relation E that contains R; and ideally we would like to know whether such an E is always unique, and, in the positive case, how to compute it. The answer is given by considering corresponding hull constructions:

- *Reflexivity:* The *reflexive hull* of R is easily obtained by adding all missing pairs of the form (a, a) where a is in A. Since sets don't care how often the same elements are added, we may define this reflexive hull of R as the union $R \cup I_A$ – recalling that I_A equals $\{(a, a) \mid a \in A\}$.

- *Symmetry:* Similarly, we may compute the *symmetric hull* of a binary relation R by adding missing pairs (b, a) for which the pair (a, b) is in R. Thus, we may define the symmetric hull of R as the union $R \cup R^{-1}$.

- *Transitivity:* The *transitive hull* (also called the *transitive closure* of relation R) is denoted by R^+ and defined to be the smallest binary relation on A that contains R and is transitive. We already say that R^+ has an inductive characterization as in (3.8). We may express this equation also in terms of elements by noting that

$$(a,b) \in R^+ \Leftrightarrow_{df} (a,b) \in R \vee \exists c \in A. \, (a,c) \in R^+ \wedge (c,b) \in R^+$$

In order to show how these three hull operations can be used to compute the smallest equivalence relation that contains R, let E be an arbitrary equivalence relation over set A that contains R. Since E is reflexive and $R \subseteq E$, we infer that

$$R \cup I_A \subseteq E \cup I_A = E$$

Since E is symmetric we can use the latter inclusion to further infer that

$$(R \cup I_A) \cup (R \cup I_A)^{-1} \subseteq E \cup E^{-1} = E$$

Let us now denote the relation $((R \cup I_A) \cup (R \cup I_A)^{-1})^+$ by $R_\equiv$. Because of the monotonicity of $^+$ and the fact that E is transitive, we infer that

$$R_\equiv \subseteq E^+ = E \tag{3.9}$$

Therefore, we learn that the relation $R_\equiv$ is contained in all equivalence relations over A that contain R. Clearly, $R_\equiv$ is transitive: it is of the form S^+ and $(S^+)^+$ equals S^+ for all binary relations S. By construction, $R_\equiv$ contains $R \cup I_A$ and so it is reflexive, since $S \subseteq S^+$ for all binary relations S. We leave it as an exercise to prove that $R_\equiv$ is also symmetric. It is actually notationally easier to prove a more general fact: let S be a symmetric binary relation over set A; then S^+ is also symmetric. But then we know that $R_\equiv$ is an equivalence relation that contains R, and therefore it must be the smallest such relation[30] – which also shows its uniqueness. We also note that the above outlines an explicit construction for computing $R_\equiv$: the hull operators for reflexivity, symmetry, and transitivity are applied in that order.

It is of interest to contemplate whether $R_\equiv$ could have been computed by applying these hull operators in a different order. Central to such considerations is that we mean to preserve the properties that the hull operators already gave us when we apply new hull operators. Adding I_A makes a relation reflexive. Since other hull operators only add more elements and don't remove any, it follows that the reflexivity will be preserved independent of the order of hull operators. Moreover, it does not matter whether we first apply the reflexive hull and then the symmetric hull, or vice versa: we leave the proof of this as an easy exercise. We also encourage the reader to see whether the transitive hull needs to be applied last or not in order to guarantee that $R_\equiv$ is being computed.

The standard algorithm for computing transitve closure is the Warshall algorithm [17]. If we start, however, with a relation that is known to be reflexive and symmetric, like in the discussion above, we can compute the transitive closure much more efficiently. Transitive closure preserves reflexivity and symmetry and there-

---

[30] We just proved that it is an equivalence relation that is contained in any equivalence relation that contains R.

fore guarantees in this case that the result will be an equivalence relation, the so-called *equivalence hull*. Exploiting Theorem 3.7 this offers a different approach to transitive closure computation that illustrates an algorithmic advantage that can often be gained by a judicious shift of representation. In Section 3.3.1, we saw how partitions induce equivalence relations and vice versa, and that this correspondence is a bijection between the set of partitions and the set of equivalence classes of a set A.[31] The alternative computation of $R_\equiv$ is quite simple and intuitive when specified at the *representational level* of partitions:

1. Initialize $P := \{\{a\} \mid a \in A\}$ as a partition in which all partition classes contain only one element.
2. For every (a_1, a_2) in R:

   a. Determine partition classes $M_1, M_2 \in P$ such that a_1 is in M_1 and a_2 is in M_2. These sets exist and are uniquely determined by P, a_1, and a_2, since P covers A and partition classes are mutually disjoint.

   b. Let $P' = (P \setminus \{M_1, M_2\}) \cup \{M_1 \cup M_2\}$ be the partition that is like P except that the two partition classes M_1 and M_2 are glued together into a single partition class for P'. It is easy to see that this process preserves the property of being a partition.

   c. Set $P \leftarrow P'$ (and go to the next element (a_1, a_2) in R in step 2 if applicable).

3. After we have done this for all pairs in R, the resulting partition P is such that its corresponding equivalence relation $\sim_P$ equals the desired $R_\equiv$.

Let us illustrate the partition-based approach by means of an example. Suppose we want to determine the smallest equivalence relation $R_\equiv$ for a relation

$$R =_{df} \{(1,3), (2,5), (6,2)\} \subseteq \{1, \ldots, 6\} \times \{1, \ldots, 6\}$$

The process starts with the initial partition $\{\{1\}, \{2\}, \{3\}, \{4\}, \{5\}, \{6\}\}$. Incorporating element $(1,3)$ of R leads to partition $\{\{1,3\}, \{2\}, \{4\}, \{5\}, \{6\}\}$, where the classes $\{1\}$ and $\{3\}$ are glued together. Using element $(2,5) \in R$ we get $\{\{1,3\}, \{2,5\}, \{4\}, \{6\}\}$, and with element $(6,2) \in R$ we finally arrive at the resulting partition $\{\{1,3\}, \{2,5,6\}, \{4\}\}$, from which we immediately conclude

$$R_\equiv = (\{1,3\} \times \{1,3\}) \cup (\{2,5,6\} \times \{2,5,6\}) \cup \{(4,4)\}$$

The assignments $P \leftarrow P'$ in step 2c) of the above algorithm *coarsen* the partition P so that each old partition class is contained in a new one. Note that this representational change from equivalence relations to partitions gives us a very efficient solution: there are at most as many partition refinement steps as there are elements in the relation R, and it is, indeed, possible to handle the refinement steps with an almost constant time bound. As a relation R over a set with n elements has at most n^2 elements, the described algorithm is almost quadratic in n. In contrast, the best

---

[31] In the second volume we will see that this correspondence is in fact a so-called lattice homomorphism, i.e., a bijection that, in this case, preserves the structure of a lattice. Homomorphisms are indeed the central notion of the second volume.

known algorithms for directly computing the transitive closure of R are essentially cubic in n, i.e., require effort which grows with n^3.

The efficient implementation of the above algorithm in a programming language is quite illustrative for the impact of the choice of data structures. Crucial for the algorithm are two operations, the matching of elements a_1 and a_2 to partition classes, and the efficient merging of two partition classes. There is a data structure called a *Disjoint Set Forest* [17], also known as the UNION-FIND data structure, which provides exactly this support. We refer to textbooks on algorithms and data structures for further details on this.[32]

Another advantage of the representational shift to partitions is that it allows us to use the form of correctness argument seen in the previous chapter on page 33: the algorithm has an **invariant** that, after **termination** at the final step, will have as a **consequence** that the final computed partition is the least equivalence relation containing R. Let us explore this a bit more formally. For two partitions P and P' over set A, we say that P is less than or equal to P', written as $P \leq P'$ if, and only if, for all partition classes C in P there is some partition class C' in P' such that $C \subseteq C'$:

$$P \leq P' =_{df} \forall C \in P. \, \exists C' \in P'. \, C \subseteq C' \tag{3.10}$$

Let P^* be the partition that corresponds to the equivalence relation $R_\equiv$, the least equivalence relation that contains relation R. We then claim that

$$P \leq P^* \tag{3.11}$$

is an **invariant** – meaning that it is true before we process the first element of R, and remains true after we process any element of R as described above. This is initially correct: the initial value of P is the partition with singleton classes $\{a\}$ so $P \leq P^*$ trivially holds. Now, consider a step in which an element (a_1, a_2) in R coarsens the partition P to P'. We have $a_1 \in M_1$ and $a_2 \in M_2$ for partition classes M_1 and M_2 of P. Since the invariant $P \leq P^*$ holds, we infer that there are partition classes C_1' and C_2' of P^* such that $M_1 \subseteq C_1'$ and $M_2 \subseteq C_2'$. Since R is contained in $R_\equiv$, we have that (a_1, a_2) is in $R_\equiv$. In terms of P^*, this means that C_1' and C_2' have to be equal, as a_i is in M_i and so in C_i' for $i = 1, 2$. But then we obtain that $M_1 \cup M_2$ is a subset of C_1'. Crucially, this ensures that

$$P' \leq P^* \tag{3.12}$$

holds, and so the update $P \leftarrow P'$ preserves the invariant in (3.11).

For the above algorithm we also have **termination** as it can repeat these refinement steps at most $|R|$ times. Finally, let P_f denote the value of P at the end of the above algorithm. We then know that $P_f \leq P^*$ by the **invariant**. We leave it as an exercise to prove the desired **consequence**, that $P^* \leq P_f$, and that this in conjunction with $P_f \leq P^*$ shows that P_f and P^* are equal.

---

[32] The computation of the least equivalence relation containing a binary relation is listed as Problem 6.7 in [6] as one of the problem formulations of a more advanced level of difficulty, and is recommended for job interviews in IT companies.

The described computation is a prime application example for *fixpoint theory* [56] which we will present in more detail in the second volume of this trilogy. Important is here that partitions, as well as equivalence relations form so-called lattices which generalize the notion of powerset by replacing intersection and union with a notion of best lower approximation (infimum) and upper approximation (supremum). The equivalence hull of a relation R can indeed be seen as the best lower approximation (in this case in fact the smallest) of all equivalence relations containing R.

Partitions constitute the basis for efficient implementations also in other contexts, most prominently for the minimization of finite automata [34, 35, 62] which can be regarded as the computation of the best upper approximation of all deterministic automata that represent the same language. The here applied principle of *partition refinement*, which is somewhat dual to UNION-FIND, is very powerful and used, e.g., to refine models in the context of verification [19, 26] and model learning [3, 72, 40] which profits from multi-dimensional versions of partition refinement [36, 71, 39].

3.4.5 Equivalence Relations in Object-Oriented Programming

Equivalence relations are of such outstanding significance that their usage often occurs subconsciously or implicitly. It is often the case that we *identify* equivalent objects, so that we treat them as exact equals which we may substitute with each other. It is worth noting that the familiar equality "=" is also an equivalence relation: its interpretation over some set A is usually the identity relation I_A.

One example of such implicit identification is offered by the programming language *Java*: its language contains a comparison operator "==", which tests whether its two arguments reference *the same object* in the heap space. Equality is now interpreted at the level of references as object identifiers, and not at the level of content or structure of objects. For example, the Java expression

$$\text{newString("foo")} == \text{newString("foo")}$$

evaluates to the Boolean value `false`, since the operator `new` guarantees the generation of a fresh object with a unique reference.

If we prefer a more *semantic* notion of equality of objects, instead of their reference identity we may use the method `equals` within class `Object`. This method is inherited by all classes a programmer defines herself and can be used for testing the structural, *semantic equality*. If the programmer wishes to overwrite this method, she needs to take care that the properties of equivalence relations are guaranteed in that overwriting implementation. We quote the relevant guidance from the official Java documentation[33] for this:

---

[33] See `http://docs.oracle.com/javase/7/docs/api/java/lang/Object.html#equals(java.lang.Object)`

" The `equals` method implements an equivalence relation on non-null object references:

- It is reflexive: for any non-`null` reference value x, `x.equals(x)` should return `true`.
- It is symmetric: for any non-`null` reference values x and y, `x.equals(y)` should return `true` if and only if `y.equals(x)` returns `true`.
- It is transitive: for any non-`null` reference values x, y, and z, if `x.equals(y)` returns `true` and `y.equals(z)` returns `true`, then `x.equals(z)` should return `true`."

3.4.6 Computing with Cardinal Numbers

We leave it as an exercise that the equipotency of sets, as defined in Section 3.2.3, is an equivalence relation in that it satisfies the formal properties of reflexivity, symmetry, and transitivity. However, we have to ask whether this is a relation over a set. Strictly speaking, there is no such set of all sets – recall our discussion of Russell's Antinomy on page 60. So we either consider this to be an equivalence relation on the class of all sets (which is not a set itself) or on some designated set of sets. The equivalence classes[34] of this relation are called *cardinal numbers*. For finite sets M, we use the notation $|M|$ for the cardinal number expressing the size of M. So we may think of elements of $\mathbb{N}$ as finite cardinal numbers in this context. It turns out that there is a smallest cardinal number amongst all infinite ones, which is denoted by $\aleph_0$. It is pronounced *Aleph-Nought*, and corresponds to the cardinality of $\mathbb{N}$. Using the Axiom of Choice it can be shown that there is a smallest cardinal number bigger than $\aleph_0$, denoted by $\aleph_1$.

The Axiom of Choice states that each family $\mathfrak{S}$ of non-empty sets has a choice function $f: \mathfrak{S} \to \bigcup \mathfrak{S}$. In other words, for each S in $\mathfrak{S}$, function f picks an element $f(S)$ in S. It is perhaps remarkable that this axiom can neither be proven nor refuted within standard set theory.

It is difficult to say anything definite about $\aleph_1$ in terms of sets that have cardinality $\aleph_1$. The Continuum Hypothesis[35] states that $\aleph_1$ equals the cardinality of the set of real numbers $\mathbb{R}$, which is the same as that of $2^{\mathbb{N}}$. However, for the standard axiomatization ZFC of Mathematics, Zermelo–Fraenkel (ZF) set theory with the Axiom of Choice, it has been shown that in the simpler set theory that removes the Axiom of Choice from ZFC set theory, the Continuum Hypothesis can neither be proved nor refuted. We then say that the Continuum Hypothesis is independent of Zermelo–Fraenkel set theory. This is a result similar to the independence result for the Whitehead problem discussed on page 56.

Based on the familiar laws of cardinality for finite sets, we can define the following algebraic operations on cardinal numbers.

---

[34] It is perhaps unfortunate that the term "classes" is used for equivalence relations; equivalence classes over a set A are of course subsets of A and so sets as well.

[35] See http://en.wikipedia.org/wiki/Continuum_hypothesis

Definition 3.17 (Operations on cardinal numbers).

1. $|A| + |B| =_{df} |A \cup B|$, if $A \cap B = \emptyset$

2. $|A| * |B| =_{df} |A \times B|$

3. $|A|^{|B|} =_{df} |A^B|$ □

The relation $=$, which is based on $\leq$,[36] which itself is based on the existence of injective functions, is a partial order on the (class of) cardinal numbers.[37] For infinite sets A and B with $A \leq B$ we have that

$$|A| + |B| = |A| * |B| = |B|$$

This is the type of interpretation of $+$ that we also saw in the *invariant* infinity $=$ infinity $+ 1$ when we discussed Hilbert's Hotel in Section 3.2.3. The algebra of *ordinal* numbers interprets $+$ and $*$ very differently though.

3.5 Learning Outcomes

The study of this chapter should give you learning outcomes as discussed below.

- A good understanding of the concept of relations, which allows you to answer confidently the following questions:

  - What is a Cartesian product?
  - How are relations defined formally?
  - Which special properties can binary relations exhibit?
  - What does it mean for a function to be injective, surjective, or bijective?

- You will also have become acquainted with the concepts and applications of cardinalities of sets:

  - What does it mean for a set to be infinite?
  - How can we compare the sizes of (infinite) sets?
  - Which infinite sets have the same cardinality, and which ones have higher cardinality than others?
  - What are diagonalization procedures?
  - Where lie the limits of countability of sets?
  - What proof idea can be used to show the uncountability of set $\{0, 1\}^{\mathbb{N}}$?

- You will also have taken in the concept of an equivalence relation:

  - Which properties define equivalence relations?

---

[36] Cf. the Theorem of Cantor-Bernstein-Schröder.

[37] We will formally define partial orders in Section 5, in Definition 5.1; the anti-symmetry is then a consequence of Theorem 3.4.

- What correspondence is there between equivalence relationships and partitions?

- You will have gained a first impression of the principle of hulls and closure operators:

  - What characterizes such hulls?
  - How can we compute such hulls?

- You will also have obtained a first impression of the *power of compositionality*:

  - The ability to generalize results for effective reuse.
  - The repeated application of generalized results.

- And you will have a good feeling for what it means that operations and functions preserve properties or structure:

  - What is being preserved, and under which circumstances?
  - An appreciation of results that concern the establishment of invariances and the preservation of structure.

3.6 Exercises

Exercise 3.1 (Image and inverse relations).
Let $R \subset A \times B$ be a relation.

1. Explain why the domain of R is the co-domain of R^{-1} and why the co-domain of R is the domain of R^{-1}.
2. Show that the image $R(a)$ of a under relation R is the same as the co-image of element a under relation R^{-1}.

Exercise 3.2 (Correctness of Proof Principle 4).
Use truth tables to show the correctness of Proof Principle 4 of Contraposition.

Exercise 3.3 (Injectivity of $f_{\mathbb{Z}}$).
Prove that the function $f_{\mathbb{Z}}$, introduced on page 87, is injective.

Exercise 3.4 (Injectivity of d).
Prove that the function d, discussed on page 88, is injective.

Exercise 3.5 (Bijectivity of function).
For the function $A \mapsto \chi_A$ defined on page 90, prove that it has $f \mapsto f^{-1}(\{1\})$ as inverse, and so is bijective.

Exercise 3.6 (Correctness of Proof Principle 3).
Use truth tables to show the correctness of Proof Principle 3 of Proof by Contradiction.

Exercise 3.7 (Proof by Contradiction). Prove that $\sqrt{2}$ is not a rational number by using the Proof Principle of Proof by Contradiction. That is to say, assume that $\sqrt{2}$ is a rational number and so of the form $\frac{n}{m}$. Then explain why you may assume that n and m have no common factor. Finally, use that fact to derive a contradiction to $\sqrt{2} = \frac{n}{m}$.

Exercise 3.8 (Proof of generalized theorem).
 Prove Theorem 3.6, a generalization of Theorem 3.5, by systematically adapting the proof steps for the latter theorem to the former. Only the case when M equals $\emptyset$ will require a bespoke argument.

Exercise 3.9 (Well-defined and bijective function).
 Prove that the function $f_\mathbb{R}$ specified on page 92 is indeed well defined, and bijective.

Exercise 3.10 (Equivalence relations).
Let $R \subseteq A \times A$ and $S \subseteq A \times A$ be two arbitrary equivalence relations on a set A. Which of the following binary relations on set A are then also equivalence relations? Either prove that this is always the case or provide a counterexample for concrete instances of A, R, and S.

1. $R \cap S$
2. $R \cup S$
3. $R; S$

Exercise 3.11 (Intersection of equivalence relations).
For $n > 1$ let $\equiv_n$ be the equivalence relation on $\mathbb{Z}$ containing all pairs (x, y) with $n \mid (x - y)$. Prove that the intersection $\equiv_3 \cap \equiv_5$ of the two equivalence relations $\equiv_3$ and $\equiv_5$ equals $\equiv_{15}$.

Exercise 3.12 (Independent properties).
Recall the properties of reflexivity, symmetry, and transitivity of a homogeneous relation $R \subset A \times A$. Show that none of these properties is implied by the other two, i.e., find appropriate choices of A and R such that, respectively

1. R is reflexive, symmetric but not transitive.
2. R is symmetric, transitive but not reflexive.
3. R is transitive, reflexive, but not symmetric.

Exercise 3.13 (Equivalence relation $\sim_\mathbb{Q}$).

1. Prove that $\sim_\mathbb{Q}$, defined on page 94, satisfies the properties (1) – (3) of Definition 3.14
2. Show that $\sim_\mathbb{Q}$ is the uniquely determined equivalence relation that identifies integer fractions that have the same numerical meaning.

Exercise 3.14 (ε-equivalent functions).

Let us consider $\mathbb{R}^{\mathbb{R}}$, the set of functions from $\mathbb{R}$ to $\mathbb{R}$, as well as a small, positive real number ε. We define two functions $f, g \in \mathbb{R}^{\mathbb{R}}$ to be ε-*equivalent*, denoted by $f \sim_\varepsilon g$, if, and only if, their input/output behavior never differs by more than ε, i.e. when

$$\forall x \in \mathbb{R}. \ |f(x) - g(x)| \leq \varepsilon.$$

Investigate whether $\sim_\varepsilon$ is an equivalence relation and carefully justify your answer.

Exercise 3.15 (Properties of equivalence classes).

Recall the definition of the equivalence class $[a]_\sim$ of an element a given on page 95.

1. Show that all elements in $[a]_\sim$ are equivalent to each other with respect to $\sim$.
2. Show that the meaning of $[a]_\sim$ is independent of the choice of representative a, that is, show that for all a and a' with $a \sim a'$ we have that the equivalence classes $[a]_\sim$ and $[a']_\sim$ are equal.

Exercise 3.16 (Transitivity of a relation).

Prove that the relation $\sim_P$, defined on page 96, is transitive.

Exercise 3.17 (Bijection between set of equivalence classes and set of partitions).

Let A be a non-empty set. Recall the mapping $\sim \mapsto P_\sim$ between the set of equivalence relations on set A and the set of partitions on A, discussed on page 97.

1. Show that this mapping is bijective and has $P \mapsto \sim_P$ as inverse function:

    a. Show that both functions are well defined.
    b. Show that $P_{\sim_P} = P$.
    c. Show that $\sim_{P_\sim} = \sim$.

2. Consider two equivalence relations R and R' on set A such that $R \subseteq R'$. Characterize the corresponding relationship between the partitions P_R and $P_{R'}$.
3. Recall the relation $\leq$ between partitions in (3.10). Show that $P \leq P'$ implies $\sim_P \subseteq \sim_{P'}$.

Exercise 3.18 (Invariant implies program correctness).

Consider the partitions P_f and P discussed on page 109. We already showed that $P_f \leq P^*$ holds since $P \leq P^*$ is an invariant and so also applies to the final version P_f of P.

1. Prove that $P^* \leq P_f$, using the Proof Principle of Proof by Contradiction.
2. Show that for any partitions P and Q over some set A, the truth of $P \leq Q$ and $Q \leq P$ implies that P equals Q. In other words, $\leq$ is *anti-symmetric*.
3. Use the previous two facts to show that P_f equals P^*.

Exercise 3.19 (Equipotency as an equivalence).

Show that the notion of equipotency, as defined in Section 3.2.3, is an equivalence relation in that it satisfies the formal properties of reflexivity, symmetry, and transitivity.

Exercise 3.20 (Properties of functions).
Which of the following functions that map from $\mathbb{Z}$ to $\mathbb{Z}$ are injective, and which ones are surjective? Carefully justify your answers.

1. $f_1(x) = 1 + x$
2. $f_2(x) = 1 + x^2$
3. $f_3(x) = 1 + x^3$
4. $f_4(x) = 1 + x^2 + x^3$

Exercise 3.21 (Proof by Contraposition).
Use the Proof Principle of Proof by Contraposition to prove the following equation for set theory:

$$A \Delta B = A \Delta C \ \Rightarrow \ B = C$$

In what sense does this equation identify an injective function?

Exercise 3.22 (Properties of functions).
Prove Theorem 3.3.

Exercise 3.23 (Cardinality of sets).
Prove that the set $\mathbb{Q} \times \mathbb{Q}$ is countably infinite.

Exercise 3.24 (Partial and total functions).
Let $f: A \hookrightarrow B$ be a partial function with domain of definition $A' \subseteq A$. That is to say, $f(a)$ is defined for all a in A' but is undefined for all a in $A \setminus A'$. Let $*$ be an element not in the set B.

1. Show that there is an equivalent representation of partial function f as a total function $f^*: A \to B \cup \{*\}$ in the sense that f uniquely determines f^* and that we can fully recover f from f^*.
2. Show that the set of partial functions from A to B is equipotent to the set of (total) functions from A to $B \cup \{*\}$.

Exercise 3.25 (Uncountability of infinite Cartesian product of countably infinite sets).
Recall the enumeration defined on page 104.

1. Define an infinite tuple $(t_i)_{i \in \mathbb{N}}$ that is in $\prod_{n \in \mathbb{N}} \mathbb{N}$ but not in that enumeration.
2. Infer from this that there are uncountably many functions from $\mathbb{N}$ to $\mathbb{N}$.
3. Use the previous facts, where appropriate, to show that the set of functions from $\mathbb{N}$ to $\{0, 1\}$ is also uncountable.

Exercise 3.26 (Tic-tac-toe game).
Recall the game of Tic-tac-toe from Section 3.4.1 on page 99.

1. We defined the configuration subsets Conf_X, Conf_O, and Conf_e for this game on page 99. Show that no pair of these subsets has disjoint intersection.
2. Give formal definitions of the sets moves_X and moves_O of moves of the respective players X and O.

3. Show that moves$_X$ and moves$_O$ are disjoint sets.
4. Define what it means for a strategy f_O of player O to be not losing.
5. Consider a strategy f_X that is not losing for player X. Do all plays that adhere to f_X end in configurations that are not completely filled?
6. Formally describe the set of configurations that can be reached from the initial configuration C^0 through legal moves of the players.
7. This exercise lets you compute non-losing strategies:

   a. Write a computer program that extracts a strategy for player X that is not losing.
   b. Do the same to extract a strategy for player O that is not losing.
   c. Generate the unique play that adheres to the two strategies you just computed, and verify that the play ends in a draw.
   d. Describe these two strategies in natural language, as algorithmic choices.

Chapter 4
Inductive Definitions

Induction: A tool for expressing *and* reasoning *about infinities through finite means.*

In Chapter 2, we learned to appreciate sets as a central mathematical structure. Sets were specified through one of the following means: either by explicitly enumerating their elements, which of course only works for finite sets, by constructing them from other sets through application of set-theoretic operations such as union and intersection, or by selecting their elements from a larger set through a predicate.

In many cases, such specification mechanisms of sets are too limited. Firstly, the structure of the elements we mean to describe and collect into sets may be arbitrarily complex, for example elements may themselves be mathematical structures such as lists, graphs, and so forth. Secondly, the use of a predicate in set comprehension presupposes that the larger set has already been constructed. For example, our definition of the set of prime numbers *Primes* on page 43 required a definition of $\mathbb{N}$, the set of natural numbers. However, we could define the latter so far only by using the "..."-notation as another means of specifying sets, as in $\mathbb{N} =_{df} \{0, 1, 2, \ldots\}$. Although we already hinted at the potential source of imprecision and ambiguity in using such notation, its use does often suggest a natural interpretation: from simple objects and some representative examples we can extract general rules for the construction of more complex objects. In the case of $\mathbb{N}$, this seems to involve only one rule that "adds one" to the current object.

In fact, the approach of constructing arbitrarily complex objects from simpler ones through the means of fixed rules is the fundamental idea of *Inductive Definitions* – the topic of this chapter. We will introduce the principle of inductive definitions first in a formalization of natural numbers, i.e., of the set of natural numbers that contains all these formal objects. This principle is of outstanding importance, since it has many areas in which it can be fruitfully applied – especially in Informatics. For example, representations of numbers, symbolic expressions, algebraic data structures, programming languages, and processor languages are – to a large extent – defined inductively. This makes such constructs not only conceptually clean; it also allows us to define algorithms, functions, and predicates inductively over such inductively defined structures. Moreover, it allows us to then reason inductively about the *correctness* of such algorithms, functions, and predicates. Therefore, induction

© Springer International Publishing AG, part of Springer Nature 2018
B. Steffen et al., *Mathematical Foundations of Advanced Informatics*,
https://doi.org/10.1007/978-3-319-68397-3_4

can be seen as an actual methodology that ranges from design and implementation to verification.

We will therefore illustrate and develop, in Chapter 5, how the inductive approach can not only be used for defining objects and structures, but how it is also central in reasoning about such objects and structures. This leads to the approach of *Inductive Proof*, which proves propositions made about inductively defined elements or sets by mirroring the inductive structure of the definitions in the proof itself. We will see that inductive definitions construct infinite sets through finitely many rules. Therefore, inductive proofs allow us to reason about infinitely many elements in such sets by examining finitely many representative cases that – in their totality – guarantee that the proof indeed applies to each of the constructed elements.

In fact, the combination of inductive definition and inductive proof is so natural and goes so hand in hand that it is without doubt one of the central methods of mastering infinite sets through finite means. In Informatics, this is especially true in the areas of programming languages and processor languages, which we will touch upon briefly in Section 4.3.

4.1 Natural Numbers

The intuitive usage of natural numbers is familiar to the reader since his or her child-hood. Although it is impossible for us as human beings to explicitly enumerate all natural numbers, we easily understand the concept of a natural number as the *result of counting the objects of a finite set*. For example, the natural number 4 is the result of counting the objects/elements in the set $\{\text{Spring}, \text{Summer}, \text{Autumn}, \text{Winter}\}$.[1]

We gain a deeper perspective in considering the aforementioned notation $\mathbb{N} = \{0, 1, 2, \ldots\}$: starting with the smallest natural number 0, we form for any natural number (for example, for the natural number 1) its corresponding next natural number (for example, the natural number 2) by adding 1 to the natural number in question. These observations form the core of the so-called *Peano Axioms*, which we will develop below. This development will also show a fundamental relationship between the concept of Inductive Definitions and the concept of Invariants, which we touched upon in the previous two chapters: the inductive definition of natural numbers guarantees, as an invariant, that the addition of 1 to a natural number preserves the property of being a natural number.

---

[1] More abstractly, we may say that the set of natural numbers is a model of all equivalence classes of $\cong$ (which expresses that sets have the same size) on finite sets. For example, the sets $\{\text{Spring}, \text{Summer}, \text{Autumn}, \text{Winter}\}$, $\{\clubsuit, \spadesuit, \heartsuit, \diamondsuit\}$, $\{\text{Earth}, \text{Wind}, \text{Fire}, \text{Air}\}$ are all equipotent. When we abstract all finite sets of four elements to their cardinality, then all these sets have the same representative, the natural number 4.

4.1.1 Peano Axioms

The **Peano Axioms** provide a formal characterization of natural numbers. Their guiding idea is that every natural number can be constructed from the smallest natural number[2] by a finite application of the *successor function* $\mathfrak{s}(\cdot)$.

Definition 4.1 (Peano Axioms).

(P_1) 0 is a natural number: $0 \in \mathbb{N}$.

(P_2) Every natural number n possesses a natural number $\mathfrak{s}(n)$ as successor:

$$\forall n \in \mathbb{N}. \, \exists m \in \mathbb{N}. \, m = \mathfrak{s}(n)$$

(P_3) 0 is not the successor of any natural number:

$$\nexists n \in \mathbb{N}. \, 0 = \mathfrak{s}(n)$$

(P_4) Different natural numbers have different successors:

$$\forall m, n \in \mathbb{N}. \, n \neq m \; \Rightarrow \; \mathfrak{s}(n) \neq \mathfrak{s}(m)$$

(P_5) *Induction axiom:* For all $M \subseteq \mathbb{N}$ with $0 \in M$ and where $n \in M$ implies $\mathfrak{s}(n) \in M$, it must be the case that M equals $\mathbb{N}$:

$$\big(\forall M \subseteq \mathbb{N}. \, 0 \in M \wedge \forall n \in \mathbb{N}. \, n \in M \Rightarrow \mathfrak{s}(n) \in M\big) \; \Rightarrow \; (M = \mathbb{N}) \qquad \square$$

The axioms (P_1) and (P_3) articulate the special role of the number zero. The axioms (P_2) and (P_4) express that the application of the successor function renders for each natural number another natural number (not zero or more than one successor, as $\mathfrak{s}(\cdot)$ is an unary function symbol), and that this function $\mathfrak{s}(\cdot)$ is injective.

Axiom (P_5) is an expression of the *Proof Principle of Induction*, which we will introduce later in Chapter 5 as Proof Principle 13. Axiom (P_5) serves two related purposes. Firstly, it guarantees a minimality requirement for the construction of set $\mathbb{N}$: any set that contains 0 and is closed under the formation of successor elements is equal to $\mathbb{N}$. Secondly, it gives us an important tool for reasoning that all natural numbers possess a claimed property $\mathscr{A}$: we define $M =_{df} \{n \in \mathbb{N} \mid \mathscr{A}(n)\}$ as the set of natural numbers that satisfy property $\mathscr{A}$, and then can prove that M equals $\mathbb{N}$ by ensuring the conditions in axiom (P_5) – that 0 is in M and that n in M implies that $\mathfrak{s}(n)$ is in M as well for arbitrary n in M.

Although most people are accustomed to natural numbers and take them for granted, in the following we will rigorously develop a theory of natural numbers

---

[2] In this trilogy, we will endorse – à la Dijkstra - the motto "counting starts at zero," and so define 0 to be a natural number, and the smallest one at that. In Mathematics, one often starts counting at 1, and so 1 is usually defined to be the smallest natural number then. In any event, Peano's axioms (P_1), (P_3) and (P_5) – which are the only ones that mention 0 explicitly – can be adapted to accommodate either of these settings.

purely relying on the five Peano Axioms. Particularly, in this setting the notion "1" denoting the natural number "one" simply refers to the Peano-based construct $\mathfrak{s}(0)$.

As a first result we derive from the Peano Axioms the following important property:

Lemma 4.1 (Existence and uniqueness of predecessor). *Every natural number n different from 0 is the successor of exactly one other natural number m. That number m is also called the* predecessor *of n.*

Proof. Let $n \in \mathbb{N}$ with $n \neq 0$. First, we show that n is the successor of some natural number. For this, it suffices to show that n is an element of the set M' defined by

$$M' =_{df} \{\mathfrak{s}(m) \mid m \in \mathbb{N}\}$$

We mean to apply axiom (P$_5$) and so we define $M =_{df} M' \cup \{0\}$ and claim that M equals $\mathbb{N}$, and so all numbers other than 0 are successors of some natural number by the definitions of M and M'. By construction, 0 is in M. Using axiom (P$_2$), we see that $m \in M$ implies $\mathfrak{s}(m) \in M$: this is so since m is in $\mathbb{N}$ and so $\mathfrak{s}(m)$ is in M' by definition of M', but M' is a subset of M. Therefore, we may invoke axiom (P$_5$) to infer that M equals $\mathbb{N}$. Since n is different from 0, this gives us $n \in M'$ as desired.

Second, the uniqueness of the predecessor follows directly from axiom (P$_4$). □

The existence and uniqueness of predecessors of positive natural numbers is crucial for the well-definedness of the inductive definitions on natural numbers that we will study throughout this chapter.

The proof of Lemma 4.1 already suggests the aforementioned Proof Principle of *Induction*, which we will explore in the next chapter in the context of a few variants of inductive proofs. To reiterate this approach: suppose we want to prove that a proposition $\mathscr{A}$ is true for all natural numbers; then we define the subset $M_{\mathscr{A}}$ of those natural numbers that satisfy proposition $\mathscr{A}$, and apply (if possible) the axiom (P$_5$) to show that $M_{\mathscr{A}} = \mathbb{N}$. As in the above proof, this requires a separate consideration of 0 (the *base case* of the induction), and of all other natural numbers (the *induction step* that assumes $m \in M_{\mathscr{A}}$ and needs to show that $\mathfrak{s}(m)$ is in $M_{\mathscr{A}}$ as well).

Limitations of Induction This approach of defining $M_{\mathscr{A}}$ won't always work, even if all natural numbers satisfy $\mathscr{A}$. The property $\mathscr{A}$ may be too complex to align it with the manner in which natural numbers are constructed by adding 1 successively. Reconsider Goldbach's Conjecture stating that every even natural number larger than 2 is the sum of two primes. We first need to adjust this different scale of "all even numbers larger than 2" to the scale of all natural numbers by defining $\mathscr{A}(n)$ to mean that $2 \cdot (n+2)$ is the sum of two primes.[a] Then for $M_{\mathscr{A}}$ as above we get that 0 is in that set since $2 \cdot (0+2) = 4$ is the sum of two primes ($4 = 2 + 2$). Now suppose that n is in $M_{\mathscr{A}}$. Then we know that there are two primes p_1 and p_2 such that $2 \cdot (n+2) = p_1 + p_2$.

In order to be able to use axiom (P5), we would have to be able to prove that $\mathscr{A}(\mathfrak{s}(n))$ holds as well. In other words, we would have to find two primes p'_1 and p'_2 such that $2 \cdot ((n+1)+2)$ is the sum of p'_1 and p'_2. The problem here is that there seems to be no inductive pattern for computing such primes p'_i from the primes p_i or any other information we have from the truth of $\mathscr{A}(n)$. Fortunately, applications of induction in Informatics very rarely run into such problems.

---

[a] We also take the liberty here to write the usual arithmetic expressions such as $2 \cdot (n+2)$ even though they would first have to be defined from Peano's Axioms.

Higher-Order Predicate Logic Let us point out here that the Induction Axiom (P5) is our first example of a formula of predicate logic of *second* order. All formulas we encountered so far had their quantifiers range over the elements of the structure, a so-called *first-order* quantification which gives rise to first-order predicate logic. If we also allow for quantifiers that range over *subsets* of structures (i.e., over relations), then we arrive at *second-order* predicates which gives rise to second-order predicate logic. This is seen here in the quantifier "$\forall M \subseteq \mathbb{N}.$" in axiom (P5). More generally, we may quantify over relations of relations (getting third-order predicates) or continue this process to obtain higher-order versions of predicate logic.

4.1.2 Operations over Natural Numbers

It is apparent that Peano's Axioms do not mention familiar arithmetic operations such as addition and multiplication. However, we may use Peano's Axioms to inductively define such operations. This not only puts such operations on a sound mathematical foundation, it also will give us a tool for proving properties of such operations using the Proof Principle of Induction. We turn to the definition of addition first.

Definition 4.2 (Addition of natural numbers). The addition of two natural numbers n and m is inductively defined as

$$0 + m =_{df} m \tag{a}$$
$$\mathfrak{s}(n) + m =_{df} \mathfrak{s}(n+m) \tag{b}$$

$\square$

Equation (a) may still be seen as a mere simplification of the expression on its left-hand side. As for equation (b), it only seems to change the position of $\mathfrak{s}(\cdot)$ on the right-hand side; but the first argument n of addition on the right-hand side is *strictly smaller* than the first argument $\mathfrak{s}(n)$ of addition on the left-hand side. This is what makes this intuitively a correct definition of a function $+: \mathbb{N} \times \mathbb{N} \to \mathbb{N}$ and the basis for intuitive computations. We illustrate the latter by computing $2+1$ using this inductive definition:

$$
\begin{aligned}
\mathfrak{s}(\mathfrak{s}(0)) + \mathfrak{s}(0) &\overset{(b)}{=} \mathfrak{s}(\mathfrak{s}(0) + \mathfrak{s}(0)) \\
&\overset{(b)}{=} \mathfrak{s}(\mathfrak{s}(0 + \mathfrak{s}(0))) \\
&\overset{(a)}{=} \mathfrak{s}(\mathfrak{s}(\mathfrak{s}(0)))
\end{aligned}
$$

This way of computing with $+$ is mathematically correct but highly impractical. But here we merely provide formal foundations for such functions and for the derivation of their algebraic properties. More efficient implementations of such functions can then appeal to these foundations to ensure that an implementation is not just efficient but also correct.

Similarly, we provide an inductive definition of multiplication next.

Definition 4.3 (Multiplication of natural numbers). The multiplication of two natural numbers n and m is inductively defined by

$$
\begin{aligned}
0 \cdot m &=_{df} 0 \\
\mathfrak{s}(n) \cdot m &=_{df} m + (n \cdot m)
\end{aligned}
\qquad \square
$$

With the aid of such inductive definitions, we may derive formalizations of the sum and product signs $\sum$ and $\prod$ (respectively) – which the reader may already be familiar with:

$$
\sum_{i=1}^{k} n_i =_{df}
\begin{cases}
0 & \text{if } k = 0 \\
\left(\sum\limits_{i=1}^{k-1} n_i \right) + n_k & \text{otherwise}
\end{cases}
$$

$$
\prod_{i=1}^{k} n_i =_{df}
\begin{cases}
1 & \text{if } k = 0 \\
\left(\prod\limits_{i=1}^{k-1} n_i \right) \cdot n_k & \text{otherwise}
\end{cases}
$$

where n_i in $\mathbb{N}$ for all i in $\{1, \ldots, k\}$.

Example 4.1 (Factorial and exponentiation). There are other important operations on the natural numbers: the *factorial function* which maps natural numbers n to their factorial $n!$, and the *exponentiation function* which maps a natural number m to its n-th power m^n. We may use the product sign defined above to inductively define these functions as well:

$$n! \; =_{df} \; \prod_{i=1}^{n} i = (\ldots(1 \cdot 2)\ldots) \cdot n$$

$$m^n \; =_{df} \; \prod_{i=1}^{n} m = \underbrace{(\ldots(m \cdot m)\ldots) \cdot m}_{n \text{ times}}$$

We should stress that these inductive definitions really should be supported with inductive proofs that these definitions are well defined. For example, our definition of $+$ is claimed to define a function $+ \colon \mathbb{N} \times \mathbb{N} \to \mathbb{N}$. So we need to ensure that for all pairs (n,m) in $\mathbb{N} \times \mathbb{N}$ there is a unique element k in $\mathbb{N}$ with $n + m = k$. Such proofs will be the subject of the next chapter.

The fact that we are able to define operations and functions inductively by appeal to Peano's Axioms means that these defined operations and functions do not really add to the expressive power of Peano Arithmetic. The latter refers to the theory of natural numbers based on Peano's Axioms. In that sense, such operations and functions are *derived* within the theory and don't bring more *expressive power* to the theory. In essence, the importance of derived operations and functions stems from the fact that they allow for a more convenient and simpler way of talking about that structure. For example, it would be very cumbersome to define the factorial function *directly* through Peano's Axioms, and too error prone for human manipulation. We saw the utility of derived expressions also in predicate logic, where predicates such as *gcd* gave us a convenient abbreviation for building and using more complex predicates.

In fact, we may think of the successive introduction of derived operations and functions as being similar to the development of software libraries, with dependencies between them. Such libraries, for example the Java API[3], allow a programmer to build Java programs by relying on packages that offer functionality from Mathematics, Graphics, Security, and so forth. And this extends the toolbox and effectiveness of a programmer considerably. Such libraries don't really increase the expressive power of the underlying programming language.[4] Nonetheless, their existence is of great practical and economic value, as their use has historically led to an increase in the number and overall quality of developed software systems. In fact, there is a dedicated research direction aiming at the development of so-called domain-specific languages based on this form of derivation.

---

[3] See https://docs.oracle.com/javase/8/docs/api/

[4] For conventional programming languages, we cannot hope to increase their expressive power: they already are what is called *Turing complete*: every computable function can be programmed in such a programming language. *Church's Thesis* (see https://en.wikipedia.org/wiki/Church's_thesis_(constructive_mathematics)) states, in simple terms, that the functions from $\mathbb{N}$ to $\mathbb{N}$ that are computable are exactly the functions from $\mathbb{N}$ to $\mathbb{N}$ that can be computed by a so-called *Turing Machine*; the reader will notice that this reads more like a definition of computability and so this hypothesis can also not be proved.

The Proof Principle of Induction (see Proof Principle 13 on page 183) traditionally uses the additive expression $n+1$ for the successor of a natural number. We
therefore record here that we are justified in doing this.

Lemma 4.2. *For all $n \in \mathbb{N}$ we have $n+1 = \mathfrak{s}(n)$.*

Proof. We define the set M that models the desired equalities as follows:

$$M =_{df} \{n \in \mathbb{N} \mid n+1 = \mathfrak{s}(n)\}$$

By the first definitional clause of $+$, we learn that 0 is in M:

$$0+1 = 0+\mathfrak{s}(0) \stackrel{(\text{Def. 4.2.a})}{=} \mathfrak{s}(0)$$

By appealing to the other definitional clause of $+$, we can show that $n \in M$ implies
$\mathfrak{s}(n) \in M$:

$$\mathfrak{s}(n)+1 \stackrel{(\text{Def. 4.2.b})}{=} \mathfrak{s}(n+1) \stackrel{(n \in M)}{=} \mathfrak{s}(\mathfrak{s}(n))$$

Thus, we may invoke axiom (P_5) and get $M = \mathbb{N}$. But this translates to the claim
made in the lemma.   □

4.1.3 Inductively Defined Algorithms

Based on the inductive definition of natural numbers, we may define a variety of
further functions that manipulate natural numbers. It is less obvious – but not less
important – to understand how to inductively define algorithms. In the Introduction,
we already got to know the problem of the *Towers of Hanoi*. There, we have n discs
with holes in them sitting on a *source* rod "S". The discs are of decreasing size and
sit on the rod such that smaller discs are on top of all larger discs (see Figure 1.2).
The problem is to move this stack of discs to the rightmost rod (called "T" for *target*
rod in the figure) so that each disc only ever sits on top of larger discs. One may
solve this by making use of an *auxiliary* rod "A". The allowed transformations are
simply described in terms of a rule (R) and a constraint (C):

(R) Only one disc may be moved at a time from any rod S, A, or T to another.
(C) None of the rods S, A, and T contain a larger disc on top of a smaller one.

These specifications leave some aspects implicit that come from the physical domain
considered here. For example, that we cannot move a disc that is not on the top of
its pile. This is physically impossible without breaking the rule (R) above. Note how
(C) above states that an *invariant* must be preserved: we cannot solve this by having
intermediate piles in which a larger disc is on top of a smaller one.

Apparently, this problem is trivial in the case of $n = 1$: we may simply move the
sole disc from the target rod to the source rod; this clearly satisfies (R) and it also
satisfies (C) as there are no other discs around that could violate that constraint.

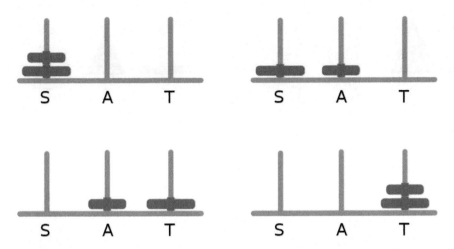

Fig. 4.1 Towers of Hanoi with two discs

The case when $n = 2$ is not that much more difficult (see Figure 4.1). In fact, the solution to that case already contains the pattern for solving the general case through an inductive algorithm: we first move the topmost disc from the source rod S to the auxiliary rod A. Then we can recognize an instance of the $n = 1$ problem in the top right corner of Figure 4.2: we may move the larger disc from the source rod S to the target rod T. This leaves us in the situation shown in the bottom left corner of Figure 4.2. Again, we recognize here an instance of the $n = 1$ problem; however, it is an instance in which we don't move from the source to the target rod but from the auxiliary rod A to the target rod T, as seen in the bottom right corner of Figure 4.2.

It turns out that it is surprisingly easy to generalize this solution from 2 to a general value of n, as suggested in Figure 4.2. We "pretend" that the $n - 1$ discs on top of pile S are just a single disc and then solve the problem as in the case for $n = 2$. Of course, this means that we now have to solve two such problems for $n - 1$ (and this is where induction will do its work). Concretely, we can describe the solution for n as a three-step process:

1. Move the topmost $n - 1$ discs from rod S to rod A.
2. Move the largest remaining disc from rod S to rod T.
3. Move the $n - 1$ discs from rod A to rod T.

The first and third step above capture a so-called *recursive descent*: a problem for n is reduced to one or more (here two) smaller but similar ones. Note how both smaller problems have their own view of source and target: in step 1 above, the target changes to A whereas the source does not change; and in step 3 above, the source changes to A whereas the target does not change.

This is almost all we need to appreciate how the recursive Algorithm 2 shown below operates. It remains to understand that the above inductive pattern already applies for the case of $n = 1$, if we imagine that the pile of $n - 1$ topmost discs is

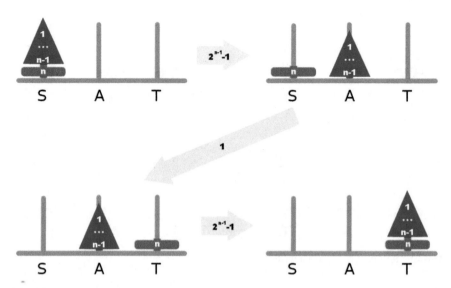

Fig. 4.2 Inductive pattern for the Towers of Hanoi

empty and so its movement won't really have any visible effect. This recursive algorithm uses a procedure *Hanoi* that has several *formal parameters*: the total number of discs n, the source rod s, the auxiliary rod a, and the target rod t. Parameter n allows us to decrease the number of discs in the recursive descent, and the remaining three parameters allow us to adjust which rod is source, auxiliary, or target in recursive calls as discussed above. The recursive call *Hanoi*$(1,'A','T','S')$, for example, models the situation in which the sole disc is on rod A and needs to be moved to rod S, and where T takes the role of the auxiliary rod.

Input: Natural number n
Output: Displacement steps that describe a solution

Hanoi(int n, char s, char a, char t) **is**
    if $n > 0$ **then**
        Hanoi$(n-1,s,t,a)$;
        Displace the topmost disc from rod s to rod t;
        Hanoi$(n-1,a,s,t)$;
end

Algorithm 2: Solving the Towers of Hanoi

The initial call *Hanoi*$(n,'S','A','T')$ will then compute the exact sequence of displacement steps needed for the solution of the Towers of Hanoi for n discs. In Figure 4.3, we see for sake of illustration the corresponding computational process for the case of $n = 3$; this is shown in the form of a so-called *call tree*.[5]

---

[5] A call tree is a schematic representation of hierarchies of executions that result from nested calls of procedures or methods. It is also used for non-recursive procedures that call other procedures.

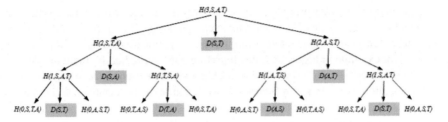

Fig. 4.3 Tree of the recursive calls of procedure *Hanoi* in Algorithm 2 for $n = 3$. In the interest of space, we abbreviated call *Hanoi(...)* to *H(...)* and indicated displacement steps through corresponding expressions *D(...)*. The displacement steps are leaves of this tree and need to be read from left to right to render a human-readable solution

A user of Algorithm 2 will only be interested in the concrete displacement steps, which we emphasized in grey in the figure. These steps, in their computed order, describe exactly how to solve the problem of the Towers of Hanoi for any value of n. All the other operations – the two recursive calls – are only organizational overhead for correctly driving that computation.

You may think that the above approach to solving this problem is very indirect. After all, we can later prove that this procedure will compute $2^n - 1$ displacement steps for solving the problem for n. However, it is possible to prove that any solution whatsoever to the problem for n will require *at least* $2^n - 1$ displacement steps. Therefore, Algorithm 2 turns out to be *optimal* for the problem of the Towers of Hanoi.

In Chapter 5 we will make use of the sketch in Figure 4.2 to formally prove that Algorithm 2 generates $2^n - 1$ displacement steps; this will use *mathematical induction*, a technique developed in that chapter. A proof that any other possible solution to this problem requires at least that many displacement steps is much harder; note that this has an implicit universal quantification over algorithms: *for every* algorithm A that solves the Towers of Hanoi, it will generate at least $2^n - 1$ displacement steps. Proving such optimality results for algorithms is typically hard and for many problems of practical importance we have algorithmic solutions but no understanding of what an optimal solution might look like. One such example is the computation of an inverse for an invertible matrix in linear algebra, a topic we will explore in volume 3 of this trilogy.

Human versus Machine The inductive definition of Algorithm 2 for the Towers of Hanoi impresses through its structural elegance. And this simplicity lends itself to transparent implementations in standard procedural or functional programming languages. Moreover, this simplicity also makes it easier to prove properties of this algorithm – such as its correctness and optimality – with the tools developed in the next chapter. However, the elegance of Algorithm 2 does not easily translate into an *operative directive* of how to concretely solve that problem. The reason resides in the two recursive calls, which are indirections for generating such operative solution plans: the concrete solution steps are found at the leaves of the corresponding call tree, as seen for example in Figure 4.3. It is therefore hard to predict, by mere inspection of this algorithm, *which* disc should be moved next onto *which* rod.

Recalling the call tree in Figure 4.3, we can at least appreciate how the first step will operate. It is obtained by traversing the call tree along its leftmost path. On this path, assignments alternate the roles of auxiliary and target rod in each step: for odd values of n (e.g., when $n = 3$), we therefore conclude that the smallest disc first has to be moved from the source to the target rod, whereas it has to be moved to the auxiliary rod for even values of n. The remaining displacement steps also follow a similar but difficult to grasp pattern, which is the foundation of corresponding iterative procedures.

The sequence of displacement steps that expresses a solution is certainly better suited as an *operative directive* for a human player. But proving correctness or optimality directly on such sequences would be more difficult than proving the same on the recursive Algorithm 2. This hints at the fact that the design of algorithms may not only depend on the problem to be solved. It may also depend on the intended usage of results computed by the algorithm. For the Towers of Hanoi, we are definitely interested in computations that humans can follow and appreciate as describing a solution of the problem. But for the multiplication of integers, for example, we are merely interested in very efficient and correct computations and not in whether the used algorithm has any resemblance to our expectations of how to multiply integers.[a]

---

[a] Some of the fastest algorithms for integer multiplication transform this problem into another one over polynomials, using discrete Fast Fourier Transforms. The program code for this approach will appear to be completely cryptic to a non-expert.

The above discussion of Algorithm 2 illustrates a crucial ingredient for the success of inductive/recursive approaches: the actual formulation of the problem often has to be suitably generalized or strengthened. In the case of the Towers of Hanoi, we can see this since the concrete task of moving n discs from rod S to rod T is not suitably inductive to allow for a solution. We had to generalize this problem in the following manner to ensure success:

- The role of a rod as being source, target, or auxiliary changes dynamically as and when needed. In other words, each rod can take on any of the three roles of source rod, auxiliary rod, or target rod.
- It is not the case that at least one rod has to be vacant. We only require that a rod is used as the target for discs that are smaller than any discs already on that rod.

The second observation also benefited from the constraint (C) on page 126, the invariant that no disc is ever on top of a smaller disc on any rod.

The original problem of the Towers of Hanoi is therefore a special case, the initial call $Hanoi(n, 'S', 'A', 'T')$, of the more general problem – the procedure

$$Hanoi(int\ n,\ char\ a,\ char\ h,\ char\ z)$$

We will see this phenomenon often in this trilogy. In fact, the determination of appropriate Archimedean Points (e.g. the invariant expressed in constraint (C)) that generalize or strengthen specified problems can be seen as fundamental in the inductive approach – as demonstrated in Section 5.7.1 below.

4.2 Inductively Defined Sets

In Section 4.1, we got to know a very simple, inductively defined structure: the set of natural numbers in its Peano Axiomatization. The number 0 was defined to be a natural number and, using that and a successor function $\mathfrak{s}(\cdot)$, we could generate *all* other natural numbers through the successive application of the successor function $\mathfrak{s}(\cdot)$ to natural number 0. Similarly to this approach, we can define more general and complex rules for the inductive definition of sets.

Definition 4.4 (Inductively defined sets). Let

1. $\mathcal{A}t$ be a set of *elementary* (also called *atomic*) building blocks
2. Op be a set of *operators* (also called *constructors*) where each operator in Op has an arity $k \geq 1$ and so can combine k smaller building blocks into a larger unit.

The set defined inductively through $\mathcal{A}t$ and Op is the smallest set $\mathcal{M}$ for which the following is true:

(C$_1$) $\mathcal{A}t \subseteq \mathcal{M}$,
(C$_2$) Let o be an operator in Op of arity k and $m_1, \ldots, m_k$ elements of $\mathcal{M}$. Then $o(m_1, \ldots, m_k)$ is also an element of $\mathcal{M}$.[6]   □

The two rules (C$_1$) and (C$_2$) define that set $\mathcal{M}$ is *closed*, we may even say *invariant*, under certain things. Rule (C$_1$) says that all atomic elements are in the inductively defined set (and so $\mathcal{M}$ is "closed" under selecting elements of $\mathcal{A}t$). Rule (C$_2$) says that the inductively defined set $\mathcal{M}$ is closed under the application of operators from Op.

Let us understand how this definition can define the set of natural numbers inductively. In that case, set $\mathcal{A}t$ is $\{0\}$, clearly expressed by Peano Axiom P_1, and set Op is $\{\mathfrak{s}(\cdot)\}$ where $\mathfrak{s}(\cdot)$ has arity $k = 1$. The Peano Axioms (P$_2$)-(P$_4$) express that the structure is "free", in the sense that there are no equations: Two expressions $\mathfrak{s}^i(0)$ and $\mathfrak{s}^j(0)$ are only considered equal if $i = j$. This condition is implicit in Definition 4.4 which, by itself, also only supports syntactical identity as equality. Finally, Peano Axiom (P$_5$) is nothing but an axiomatic specification of *smallest*, a general and explicitly stated requirement for all structures defined inductively according to Definition 4.4.

---

[6] A more compact but less intuitive definition of inductively defined sets allows for operators of arity 0, which can then represent elements of the set $\mathcal{A}t$ and so only item (C$_2$) above is needed then.

In fact, this requirement poses the question whether Definition 4.4 is actually well defined. Hypothetically, it could be the case that there does not exist a *smallest* set that is closed under rules (C_1) and (C_2). Luckily, its existence can always be guaranteed, indeed with an argument similar to the existence of hulls as discussed in the previous chapter. We will explore the reasons for this in the exercises as well as in the second volume of this trilogy.

In general, the operators in *Op* may have arbitrary arity: for example, negation has arity 1 whereas conjunction has arity 2 in the set of Boolean terms inductively defined below, and if we would like to establish a scenario with the function *Hanoi* as a dedicated operator, its arity would be 4.

4.2.1 Applications of Inductive Definitions in Informatics

Inductive definitions are a concept and tool that permeate Informatics. Logics, terms and expressions, programming languages, calculi, and data structures are all typically defined inductively. We now illustrate this through a few examples.

Example 4.2 (Binary Trees). The set of binary trees is the smallest set closed under the following rules:

1. The *empty binary tree* ∘ is an atomic binary tree.
2. Whenever T_1 and T_2 are binary trees, then $\bullet(T_1, T_2)$ is also a binary tree that has T_1 as left sub-tree and T_2 as right sub-tree.

The inductively constructed binary tree

$$\bullet(\bullet(\bullet(\circ, \circ), \bullet(\bullet(\circ, \circ), \circ)), \bullet(\circ, \circ))$$

is an example of a binary tree. We can better understand its structure if we depict it graphically as seen in Figure 4.4. Such tree representations are typical in Informatics: the atomic expressions are the leaves and shown at the bottom (in contrast to what one finds in nature); and the construction rules combine sub-trees and/or leaves into larger sub-trees.[7]

Operators as Constructors We want to emphasize the separation of syntax and semantics, which is central in Informatics. In a syntactic view, we consider constructs $o(m_1, \ldots, m_k)$ built out of elements m_i from a set $\mathcal{M}$ and through operators o as mere syntactic expressions, which still require a formal semantic interpretation (see Section 4.3) before we think of them as semantic objects. In (Pure) Mathematics, such a strict separation of syntax and semantics is often not made, resulting in a semantic conception of inductive sets:

---

[7] The representation of inductively defined expressions as trees has been established for practical reasons: it aids the intuitive illustration of such structures and it is a natural data representation for computations performed over such expressions/trees.

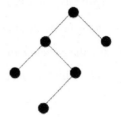

Fig. 4.4 Graphical representation of the binary tree $\bullet(\bullet(\bullet(\circ,\circ),\bullet(\bullet(\circ,\circ),\circ)),\bullet(\circ,\circ))$ where atomic leaves $\circ$ are not shown

operators op from set the Op are then considered k-ary functions. The corresponding formalism will then require a *carrier set* $G \supseteq M$, so that we may define these functions op without explicit reference to M.

For example, suppose that we have defined the set of natural numbers $\mathbb{N}$. Then we may define the set of natural numbers that are a power of 2, set P, by setting $Op =_{df} \{d\}$ and defining d as the doubling function $d : \mathbb{N} \to \mathbb{N}, d(n) = 2 \cdot n$. We may now define P inductively, by stating the set of atomic elements as $At =_{df} \{1\}$ and by having just one closure rule saying that the set P is closed under applications of d. In contrast to Definition 4.4, expressions of the form $d(d(\cdots d(1)\cdots))$ are then merely a guide for how to compute their meaning, the elements of P, as set $M =_{df} P$ was inductively defined *over the ground set* $G =_{df} \mathbb{N}$.

In contrast, in Informatics it is often better to follow the syntactic view and to construe operators as *constructors*. The construct $o(m_1, \ldots, m_k)$ is then not a recipe for a computation – informed by an interpretation of o as a function on a set. Rather, it describes the inherent structure of the respective type of object, for example that of a particular binary tree. In this approach, it is not required to specify a ground set, as the inductive definition is the mechanism that introduces, in the first place, the expressions that lead to such sets – for example, the set of binary trees. We will follow this approach in the remainder of this trilogy.

Boolean Terms

Another important example of an inductively defined set is that of *Boolean terms*. Recall that we already defined formulas of propositional logic in Section 2.1. Boolean terms are nothing but an inductively defined and so more formal account of such formulas.

Definition 4.5 (Boolean terms). Let $\mathcal{V}$ be a set of Boolean variables, e.g., $\mathcal{V} = \{X, Y, Z, \ldots\}$. The set $\mathcal{BT}$ of all *Boolean terms* over $\mathcal{V}$ is the smallest set closed under:

1. T, F, as well as Boolean variables from set $\mathscr{V}$ are atomic Boolean terms.

2. Let t_1 and t_2 be Boolean terms. Then the following are also Boolean terms:

   - $(\neg t_1)$, the *Negation* of t_1,

   - $(t_1 \wedge t_2)$, the *Conjunction* of t_1 and t_2, and

   - $(t_1 \vee t_2)$, the *Disjunction* of t_1 and t_2. □

We emphasize that the above definition describes only the *syntax*, that is to say the exterior form, of Boolean terms. In doing so, we do not yet ascribe any meaning to the operators such as $\vee$ and $\wedge$. For example, such a meaning was articulated in natural language in Definition 2.2 on page 23. In Section 4.4 we will assign a *semantics* to Boolean terms; this will be done formally through an inductive definition, not in natural language. Therefore, such an approach is also called a *formal semantics*. This inductive definition is then a recipe for how to compute the truth value of a Boolean term, given truth values for all its Boolean variables, in a way that can be directly implemented.

Before we do this, we will provide another example of an inductive definition that formalizes an important syntactic operation on Boolean terms: substitution.

Definition 4.6 (Syntactic substitution). Let t, t_1, and t_2 in $\mathscr{BT}$ be Boolean terms and X, Y in $\mathscr{V}$ variables. Then we define the *syntactic substitution* $\cdot[\cdot \mapsto \cdot] : \mathscr{BT} \times \mathscr{V} \times \mathscr{BT} \to \mathscr{BT}$ inductively as follows:

- $\mathsf{T}[X \mapsto t] =_{df} \mathsf{T}$

- $\mathsf{F}[X \mapsto t] =_{df} \mathsf{F}$

- $Y[X \mapsto t] =_{df} \begin{cases} t & \text{if } Y = X \\ Y & \text{otherwise} \end{cases}$

- $(\neg t_1)[X \mapsto t] =_{df} (\neg t_1[X \mapsto t])$

- $(t_1 \wedge t_2)[X \mapsto t] =_{df} (t_1[X \mapsto t] \wedge t_2[X \mapsto t])$

- $(t_1 \vee t_2)[X \mapsto t] =_{df} (t_1[X \mapsto t] \vee t_2[X \mapsto t])$ □

Intuitively, expression $s[X \mapsto t]$ is the Boolean term that results from replacing all occurrences of variable X in Boolean term s with Boolean term t. The induction is over the structure of s, whereas t is an arbitrary but fixed Boolean term.

What is remarkable in the above definition is the fact that the substitution operator $\cdot[X \mapsto t]$ is applied always in the same manner to non-atomic Boolean terms: the exterior form of the argument term does not change; negations remain negations, conjunctions remain conjunctions, and so forth. In this context, we say that the substitution operator is *compositional*. This means that its effect may be computed from its effects on smaller sub-terms. Inductive definition can be regarded as a best practice for the design of languages that support modular design, which, in particular, comprises the inductive definition of the corresponding semantics (see page 147). The general, underlying mathematical principle for this concept is that of a *homomorphism*, which is a central theme in the second volume of this trilogy, *Algebraic Thinking*.

Example 4.3 illustrates the application of the substitution operator. In the case of the first substitution, it shows all intermediate steps of the computation according to the inductive definition of that operator. For the second substitution, we leave it as an exercise to make the intermediate steps explicit.

Example 4.3.
- $(\neg(Y \wedge X))[X \mapsto t] = (\neg(Y \wedge X)[X \mapsto t]) = (\neg(Y[X \mapsto t] \wedge X[X \mapsto t])) = (\neg(Y \wedge t))$

- $(X \vee (Y \wedge X))[X \mapsto t] = \cdots = (t \vee (Y \wedge t))$

As already mentioned, inductive definitions are a central tool in Informatics. We will encounter and investigate this tool more closely in this trilogy. For example, we will study the application of inductive definitions to the construct of *lists* in functional programming languages in Section 4.5.

4.3 Representation and Meaning

The fundamental role of Informatics is the systematic processing of information, especially its automated processing through the aid of computing engines. This includes the development of such engines and means of processing – where conflicting aspects such as energy efficiency, performance, and security may constrain all this. The concept of *information* is here understood as capturing abstractly the meaning of a concept from or related to the real world. Such information is typically ambiguous unless we use context or additional information to clarify the intended meaning. For example, "21:10" is information but it may refer to a number of things; say, the final score in a set of table tennis, the time of day, an arithmetic expression involving division, or a reference to a verse in the Book of Job in the Bible. Information processing therefore needs to ensure that such ambiguities are resolved, and failure to do so can result in security or safety violations.[8]

The communication and subsequent processing of information requires a schematic, formalized *representation*. Dual to the situation of ambiguity, where the same representation such as "21:10" may refer to different things, the same information may be represented in different ways. At first thought, this seems problematic but we can often exploit different representations of the same information to good advantage, for example for optimizations of coding schemes to improve communication speed. Here are some examples of the usual representations of the natural number "four":

---

[8] See http://sunnyday.mit.edu/accidents/Ariane5accidentreport.html

- Decimal: **4**

- Binary: **100**

- Unary: **I I I I**

- Roman: **IV**

Again, we can see that the same representation may mean different things. For example, expression **IV** may be an acronym (as a sequence of two letters of the alphabet) and so stand for many things ranging from an activist organization, a periodical, and a postcode area to an unincorporated community near Santa Barbara in California[9]. Oh, and it may also stand for the number "four". Representational bias lets us resolve such choices, so Mathematicians and Informaticians will most likely pick the latter interpretation given no additional context.

The grounding of a suitable representational system (a *language*), together with a corresponding and fitting concept formation is a central task in Informatics, which we will define as *Semantic Schemas* in Section 4.3.2. The *Interpretation* then renders for each representation its *Semantics* (meaning). Without an interpretation, all representations are void of any meaning. Only the association of meaning to representations turns the latter into information. We refer back to Section 1.2 of the introduction for additional intuitive references to that topic. An informatician therefore does not want to be in the role of a Criminal Investigator, who may have lots of representations as evidence but seeks interpretations that provide the needed meaning for solving a criminal case.

In daily life, we often do not differentiate between information and its representation. Rather, we most often assume, implicitly, a *standard interpretation*. But even that may be a function of cultural norms. For example, the color red stands for mourning in southern Africa but stands for success or triumph for the Cherokees. In Informatics, there is no such standard interpretation. This gives us great flexibility for the representation and processing of information. But it also gives us great responsibility for properly separating the abstract information content from its exterior, syntactic form. The following subsections present the foundations for a "responsible" engineering of representation.

4.3.1 Character Strings

Usually, syntactic representations are sequences of symbols drawn from an alphabet. Such sequences are called *strings* or *finite words*. Mathematically, we may consider sequences of symbols of length n as functions that map from the set $\{1, \ldots, n\}$ to some alphabet A.[10]

---

[9] See http://en.wikipedia.org/wiki/IV

[10] This results in another perception of n-tuples, i.e., elements of the n-fold Cartesian product. Infinite character strings (or infinite words), which will not be considered here but which are important in various fields in Informatics, can be modeled as functions of type $\mathbb{N} \to A$.

Definition 4.7 (Strings or finite words). Let A be a finite set of symbols, also referred to as an *alphabet*. A *string* (also called a *finite word*) w of length n in $\mathbb{N}$ over alphabet A is a function

$$w : \{1, \ldots, n\} \to A$$

For $n = 0$ we consider $\{1, \ldots, n\}$ to be the empty set. The unique string of length 0 is also called the *empty word* and denoted by ε.   □

We write A^n to denote the set of all strings of length n over alphabet A. In particular, we have that A^0 equals $\{\varepsilon\}$. The set of all strings of arbitrary but finite length over alphabet A is denoted by A^* and called the *Kleene hull* of alphabet A.[11] We have that

$$A^* =_{df} \bigcup_{n \in \mathbb{N}} A^n \tag{4.1}$$

It is apparent that A^* contains also the empty word ε. If we exclude this case, we arrive at the set of non-empty finite words, denoted by A^+ and defined by

$$A^+ =_{df} \bigcup_{n \in \mathbb{N}_{>0}} A^n$$

Put differently, we have that A^+ equals $A^* \setminus \{\varepsilon\}$ An element w in A^* has a uniquely determined length, which we normally denote by $|w|$. An important binary operation on finite words is *concatenation*, which we may think of as the sequencing of two character strings.

Definition 4.8 (Concatenation of strings/finite words). Let w_1 and w_2 be character strings of length n, respectively, m over A. Then the *concatenation* of w_1 and w_2 is defined as follows:

$$w_1 w_2 : \{1, \ldots, n+m\} \to A$$

$$(w_1 w_2)(i) = \begin{cases} w_1(i) & \text{if } 1 \leq i \leq n \\ w_2(i-n) & \text{if } n+1 \leq i \leq n+m \end{cases}$$

□

It is pretty obvious that $|vw| = |v| + |w|$ holds for arbitrary elements v and w in A^*. This can also be expressed by saying that the length function $|\cdot| : A^* \to \mathbb{N}$ is compatible with the operation of concatenation on A^* and addition on $\mathbb{N}$. This compatibility is a special form of *homomorphism*, where the effect of the concatenation operator concerning the length of words is nothing but the sum of the lengths of the argument words.

---

[11] The "Kleene *" is a perfect example of a well-established finite representation of an infinite artifact.

This illustrates that the length function establishes some structural similarity between its domain and its codomain. Remember that the substitution operator on Boolean terms is also a homomorphism. We will see that this is, in fact, a direct consequence of the pattern of its definition: the substitution operator evaluates each conjunct of a conjunction and then forms the conjunction of the obtained results.

For the substitution operator, the domain and co-domain are the same. In contrast, the length function's domain and codomain are very different, as are the operations on them that are compatible. Function $|\cdot| : A^* \to \mathbb{N}$ is surjective for a non-empty alphabet A: for all n in $\mathbb{N}$ there are words of length n. The role of ε in A^* is similar to the role of 0 in $\mathbb{N}$: the former does not change a finite word if concatenated to it, the latter does not change a natural number if added to it. And $|\cdot|$ maps ε to 0; in fact, $|\cdot|$ is a homomorphism of *monoids* – an algebraic structure we will study in the second volume of this trilogy.

We can generalize the concatenation of finite words to sets of finite words. Such sets are an instance of so-called *formal languages*. For example, for alphabet $A = \{a,b\}$ the set of finite words whose last (rightmost) character equals a is such a formal language, and an infinite set. Concretely, let W_1 and W_2 be sets of finite words over a common alphabet A. That is to say, W_1 and W_2 are subsets of A^*. Then, we define $W_1 W_2$ (or for the sake of clarity also written as $W_1 \cdot W_2$) as the set

$$W_1 \cdot W_2 =_{df} \{w_1 w_2 \mid w_1 \in W_1, w_2 \in W_2\}$$

of finite words over A obtained by concatenation of any finite word of W_1 with any finite word of W_2. For example, let A be $\{a,b\}$ as before where W_1 is the set of finite words over A that end in character a and W_2 the set of finite words over A that end in character b. We leave it as an easy exercise to show that then

$$W_1 \cdot W_2 = \{w \in A^+ \mid w_{|w|} = b \wedge \exists i.\ 1 \le i < |w| \wedge w_i = a\} \qquad (4.2)$$

Furthermore, we denote by W^n, for n in $\mathbb{N}$, the set of finite words that is obtainable by the n-ary concatenation of finite words from W. We may define this family of sets inductively as follows:

$$W^0 =_{df} \{\varepsilon\}$$
$$W^n =_{df} W \cdot W^{n-1} \quad \text{for } n \in \mathbb{N}_{>0}$$

Similarly to the definitions given for alphabets, we may definite the Kleene hulls of a formal language W – a subset of A^* – as follows:

$$W^* =_{df} \bigcup_{n \in \mathbb{N}} W^n \qquad\qquad W^+ =_{df} \bigcup_{n \in \mathbb{N}_{>0}} W^n \qquad (4.3)$$

If we interpret elements of set A as finite words of length 1, then $W = A$ is a subset of A^*. But now we have two possible definitions of A^*, that in (4.1) and that in (4.3). We leave it as an exercise to show that the two definitions (and similarly the two definitions for A^+) coincide as long as ε is not an element of the underlying alphabet A – which is normally the case.

4.3.2 Semantic Schemas

The concept of *semantic schema* formally captures the connection between information and representation.

Definition 4.9 (Semantic schema). A *Semantic Schema* is a triple $(\mathscr{R}, \mathscr{I}, [\![\cdot]\!])$ where

- $\mathscr{R}$ is the *Set of Representations*,

- $\mathscr{I}$ is the *Set of Information*, and

- $[\![\cdot]\!] : \mathscr{R} \to \mathscr{I}$ is a (partial) function, a *Semantic Interpretation*.   □

Semantic interpretations are defined as (partial) functions that map representations to information.[12]

But in the context of natural languages, we typically deal with non-functional semantic relations. For example, in English the word "crane" may convey information that depends on context: a mechanical lifting machine, or a large bird with long legs and long neck. False cognates are finite words that have, sometimes dramatically, different meanings across two natural languages. For example "chair" in French can refer to flesh whereas in English it normally refers to a piece of furniture and definitely not to flesh. This illustrates that it is also important to choose the right semantic schema. Our definition can accommodate such situations well, since $\mathscr{I}$ may be a power set so that a representation r in $\mathscr{R}$ gets mapped to a set $[\![r]\!]$ of information.

The following semantic schemas for the representation of natural numbers illustrate the difference between information and its representation. We begin with the most simple representation, its unary one obtained by having an alphabet with a sole character.[13]

Example 4.4 (Unary representation of positive natural numbers).

- $\mathscr{R}_u =_{df} \{\mathbf{I}\}^+ = \{\mathbf{I}, \mathbf{II}, \mathbf{III}, \ldots\}$

---

[12] Partial functions are often used as a natural means whenever some syntactic constructs do not have a reasonable semantic interpretation. For instance, the effect of a non-terminating loop in an imperative programming language could be modeled by leaving it undefined, i.e., $[\![\mathbf{while}\ (true)\ \mathbf{do\ skip\ end}]\!] = undef$. On the other hand, partial semantic functions can always be rendered total by adding an explicit element representing undefinedness to the information set I.

[13] In order to clearly separate representational elements and informational elements, we will for now express elements of the representations in **boldface**.

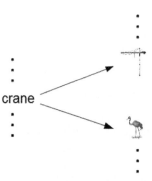

Fig. 4.5 A semantic relation
that is not right unique and so
not a function

- $\mathscr{I}_u =_{df} \mathbb{N}_{>0} = \{1, 2, \ldots\}$ where the latter is the set of positive natural numbers as information or concept, and not as their representation in the decimal system!

- $[\![\cdot]\!]_u$ is defined by $[\![\underbrace{\textbf{I I} \ldots \textbf{I}}_{n \text{ times}}]\!]_u =_{df} n$

The semantics of a message consisting of n bars **I** is therefore the natural number n.

In daily usage of natural numbers, we use their decimal representation.

Example 4.5 (Decimal representation of natural numbers).

- $\mathscr{R}_d =_{df} \{\textbf{0}, \ldots, \textbf{9}\}^+$

- $\mathscr{I}_d =_{df} \mathbb{N}$

- $[\![\cdot]\!]_d$ is defined by $[\![w]\!]_d =_{df} \sum_{i=1}^{n} 10^{n-i} \cdot [\![w(i)]\!]_z$

  In the definitions above, $[\![\cdot]\!]_z \colon \{\textbf{0}, \ldots, \textbf{9}\} \to \mathbb{N}$ denotes the value of a decimal digit, meaning that $[\![\textbf{0}]\!]_z =_{df} 0,$
  $$\vdots$$
  $[\![\textbf{9}]\!]_z =_{df} 9.$

The representation from Definition 4.5 is not very concise as $[\![\cdot]\!]_d$ is not injective. The same number can be represented, in fact, in infinitely many ways. The problem resides in the possibility of leading zeros, which allow for the following equalities:

$$[\![\textbf{1}]\!]_d = [\![\textbf{01}]\!]_d = [\![\textbf{001}]\!]_d = \ldots = 1$$

Leading zeros can be avoided by constraining the set of representations as follows:

$$\mathscr{R}_d =_{df} \{\textbf{0}\} \cup \{zw \mid z \in \{\textbf{1}, \ldots, \textbf{9}\}, w \in \{\textbf{0}, \ldots, \textbf{9}\}^*\} \tag{4.4}$$

Therefore, we solve this problem in the same manner as we solved a corresponding problem in the representation of positive rational numbers as fractions of natural numbers (recall Section 3.2.3): by adapting a *canonical* set of representations. This is a common technique when choosing a suitable representational structure.

Information and representation in Cryptography Sometimes, such uniqueness comes at a price. For example, in cryptography we may have to pad a message so that its length is a multiple of the block size of an encryption algorithm. But padding has to be done in a manner that allows us to recover the original unpadded message from the padded one; yet, one can show that all such unique padding schemes require that the padded version of some unpadded messages has one more block than the unpadded message. This is hardly an option in applications in which the messages are always short and where padding would therefore markedly increase network traffic.

This example also illustrates that information and representation can be very close to each other. In this setting, the set of information is the set of unpadded messages, and the set of representations is the set of padded messages for a given padding scheme. It is perhaps no accident that the cryptographic literature really does call these padding *schemes*.

We conclude our discussion of representations of natural numbers by studying the binary representation of numbers, a very important representation in Informatics.

Example 4.6 (Binary representation of natural numbers).

- $\mathscr{R}_b =_{df} \{0\} \cup \{1\,w \mid w \in \{0,1\}^*\}$

- $\mathscr{I}_b =_{df} \mathbb{N}$

- $[\![\cdot]\!]_b$ is defined by $[\![w]\!]_b =_{df} \sum_{i=1}^{n} 2^{n-i} \cdot [\![w(i)]\!]_{bz}$

  In the above, expression $[\![\cdot]\!]_{bz} \colon \{0,1\} \to \mathbb{N}$ denotes the value of a binary digit, meaning that $[\![0]\!]_{bz} =_{df} 0$ and $[\![1]\!]_{bz} =_{df} 1$.

We already saw that representations may refer to different pieces of information, by varying the used semantic schema. This is quite interesting for the binary representation of natural numbers. We may interpret the latter as finite sets of natural numbers, rather than as a single natural number. A finite word over $\{0,1\}$ is then seen as a bitvector that encodes a characteristic function (see Section 3.4.2): the symbol 1 then represents a present element, whereas symbol 0 represents an absent element. Given that we work with finite binary words with no leading zeros, it makes sense to represent the largest natural number through the leading 1 of the finite binary word. For example, we would then expect that $[\![1011]\!]_{bs} = \{0,1,3\}$.

Example 4.7 (Binary representation of finite sets of natural numbers).

- $\mathscr{R}_{bs} =_{df} \{0\} \cup \{1\,w \mid w \in \{0,1\}^*\}$

- $\mathscr{I}_{bs} =_{df} \mathfrak{P}(\mathbb{N})$

- $[\![\cdot]\!]_{bs}$ is defined by $[\![w]\!]_{bs} =_{df} \{ |w| - i \mid i \in \{1, \ldots, |w|\} \wedge w(i) = 1 \}$

We leave it as an exercise to show that $[\![\cdot]\!]_{bs} : \mathscr{R}_{bs} \to \mathfrak{P}(\mathbb{N})$ is injective but not surjective – noting that we cannot represent infinite subsets of $\mathbb{N}$ through finite binary words. We refer to Section 4.4 for a more general discussion of syntactic schemas in the context of inductive definitions.

4.3.3 Backus–Naur Form

The character strings that we encountered in the semantic schemas above were, admittedly, very simple in nature. But syntactic structures are often more complex in their construction. This complexity can typically be tamed by using a special format for the inductive definition of syntactic structures (i.e., formal languages): the so-called *Backus–Naur Form* (BNF).

A BNF consists of finitely many rules of the form <N> ::= w. Such rules may also be referred to as *production* rules (to emphasize the computational aspect) or *derivation* rules (to emphasize a proof-theoretic aspect). The left-hand sides of rules are formed out of so-called *non-terminal symbols*, whereas the right-hand sides of rules consist of finite (possibly empty) words that may contain non-terminal symbols as well as *terminal symbols*. You may think of the terminal symbols as the letters of the considered alphabet A. Terminal symbols don't occur as left-hand sides of rules and so cannot be used to derive finite words. A BNF can now be considered as a means to generate finite words over the alphabet of terminal symbols, and the sole role of non-terminal symbols is to facilitate this process as triggers of rules. In order to distinguish terminal from non-terminal symbols, we will put all non-terminal symbols into angle brackets. If the same left-hand side occurs in more than one rule, as in

$$\text{<N> ::= } w_1$$
$$\cdots$$
$$\text{<N> ::= } w_n$$

then we may group them in more readable form as in

$$\text{<N> ::= } w_1 \mid \ldots \mid w_n$$

Let us consider a very simple example of a BNF.

Example 4.8 (BNF for natural numbers). Based on the formalism of BNFs and aided by the Peano Axioms, we may reduce the construction of natural numbers to the following two rules with identical left-hand sides:

$$\text{<Nat> ::= } 0 \mid \mathsf{s}(\text{<Nat>}) \tag{4.5}$$

In these rules, the symbols "0","s", "(", and ")" are terminals, whereas <Nat> is the sole non-terminal symbol.

This single line of "BNF code" defines a language that syntactically represents the natural numbers and reflects faithfully all five Peano Axioms.

How can this be? The first Peano Axiom requires 0 to be a natural number and is explicitly covered. So is the required existence of a (unique) successor $\mathsf{s}(n)$ for each natural number n (the second Peano Axiom). The other three Peano Axioms are consequences of two essential conventions of BNFs:

- the syntactic (free) interpretation[14] of terms, that two different strings also mean different things, and
- the minimality requirement of the sets defined via BNF, i.e., everything must be constructible in finitely many steps by applying the BNF rules.

In particular, the fifth Peano Axiom, the foundation for natural induction, is nothing more than an elegant formulation of the minimality requirement.

Thus, in contrast to the Peano Axioms, which specify natural numbers from scratch, the BNF formulation is based on two powerful conventions: the term interpretation and the minimality requirement. It is the power of such conventions that provides the leverage in the resulting domain-specific scenario.

Parser Generation Automatic parser generation [2, 49, 4] is an extreme case of such a domain-specific scenario. In fact, BNF specifications are sufficient to entirely generate the corresponding parser code [41, 16, 63]. This impressively shows the leverage of the distinction between description in terms of WHAT and HOW: the BNF describes only the syntax of the envisioned language (the WHAT), whereas the parser generated from it is a complex program (the HOW) that automatically reads a string from a file, tokenizes it, and builds an abstract syntax tree (AST), all along checking for syntax correctness. The impact of this leverage depends on domain knowledge about parser generation, and goes well beyond what is attainable with what we traditionally would call code generation.

To build up intuition, let us now study a slightly more complex example of a BNF.

Example 4.9. Consider the BNF with two rules

$$<N> ::= a \mid a<N> \tag{4.6}$$
$$<M> ::= bb \mid bb<M> \mid b<N>$$

where a and b are the terminal symbols and <N> and <M> are the non-terminal symbols. It should be intuitively clear that the first rule can generate any finite word of form $a \cdots a$ of length $k \geq 1$. As for the second rule, it is more complex. We notice

---

[14] One sometimes speaks of the term or Herbrand interpretation.

that the second rule may either call itself recursively, or it may call the first rule; and the first rule can never call the second rule. From that we learn that <M> can generate finite words $b \cdots b$ of even length ≥ 2, or finite words $b \cdots b a \cdots a$ consisting of an odd number of bs followed by a positive, finite number of as. In the next chapter, we will see how we can inductively prove which set of finite words a BNF generates.

BNFs provide a *formal method* for the description of formal languages, which we understand to be specific sets of finite words over a finite alphabet. In the last example, the BNF seemed to describe two such formal languages. In practice, a BNF will have a designated non-terminal symbol as *start symbol* – let us say <Nat> for the BNF in (4.5). We can then apply any of the rules for that start symbol to arrive at a new finite word over the set of terminal and non-terminal symbols. In such a finite word, any non-terminal symbol may be replaced – in place – by the right-hand side of a matching rule. And this process can continue for as long as there are non-terminal symbols in the finite word. For example, we may derive a syntactic representation of the natural number 3 as follows:

$$
\begin{aligned}
\text{<Nat>} &\Longrightarrow \mathfrak{s}(\text{<Nat>}) \\
&\Longrightarrow \mathfrak{s}(\mathfrak{s}(\text{<Nat>})) \\
&\Longrightarrow \mathfrak{s}(\mathfrak{s}(\mathfrak{s}(\text{<Nat>}))) \\
&\Longrightarrow \mathfrak{s}(\mathfrak{s}(\mathfrak{s}(0)))
\end{aligned}
$$

We may formalize the notion of derivation by defining a *derivation relation* $\Longrightarrow$. Let $\mathbf{T}$ be the set of terminal symbols, $\mathbf{N}$ the set of non-terminal symbols, and $\mathbf{R}$ the set of rules of a BNF. Then the derivation relation $\Longrightarrow \subseteq (\mathbf{N} \cup \mathbf{T})^* \times (\mathbf{N} \cup \mathbf{T})^*$ is defined as follows:[15]

$$
\begin{aligned}
v \;\Longrightarrow\; w \;&\Leftrightarrow_{df} \\
&\exists v_1, v_2 \in (\mathbf{N} \cup \mathbf{T})^*.\; \exists\, (\text{<A>} ::= u) \in \mathbf{R}.\; v = v_1 \text{<A>} v_2 \wedge w = v_1 u v_2
\end{aligned}
$$

In a derivation step, we replace in a finite word v over terminal and non-terminal symbols an occurrence of non-terminal <A> with the finite word u, provided that <A> $::= u$ is a rule of the BNF. In particular, the parts of v to the left and right of <A> remain unaffected; the rule is applied "in place" or "in context". We may abbreviate the above definition of a derivation step as

$$
v_1 \text{<A>} v_2 \Longrightarrow v_1 u v_2
$$

where the existential choice of a rule has already been made. A sequence of finite words $w_1, \ldots, w_k$ over alphabet $(\mathbf{N} \cup \mathbf{T})^*$ (i.e., $w_1, \ldots, w_k \in (\mathbf{N} \cup \mathbf{T})^*$) is called a *derivation sequence*, if we have $w_i \Longrightarrow w_{i+1}$ for all i in $\{1, \ldots, k-1\}$. Furthermore,

---

[15] Here we encounter another example of overloading of notation. The derivation relation and the implication operator of logic are denoted by the same symbol. The use context of this symbol therefore needs to provide its exact meaning (recall Section 4.5.1).

we say that the finite word $w' \in (\mathbf{N} \cup \mathbf{T})^*$ is *derivable* from a finite word w in $(\mathbf{N} \cup \mathbf{T})^*$ if there is such a derivation sequence $w_1, \ldots, w_k$ with $w = w_1$ and $w' = w_k$.

For a given non-terminal symbol <A> in a BNF, this defines the *language generated by* <A> *in that BNF* as the set of finite words w over $\mathbf{T}$ (i.e., w in $\mathbf{T}^*$), that are derivable from <A>. Please note that the finite words generated by <A> must not contain any non-terminals by definition. Let us consider an example. Subsequently, let us write c^k for a word of length $k \geq 0$ consisting only of the letter c, where c^0 is understood to equal ε.

Example 4.10. For the BNF in (4.6), the language generated by <M> equals $\{b^{2k} \mid k \geq 1\} \cup \{b^{2k+1}a^l \mid k \geq 0, l \geq 1\}$. We leave the proof of this as an exercise.

BNFs are an object of study in their own right, not just a tool for elegantly defining inductive syntactic structures. *Formal Languages*, an area in Theoretical Informatics, subsumes the study of BNFs. Modern applications of formal languages range from program analysis and the modeling of distributed systems to complexity theory and the security of programming languages. But formal languages also have applications outside of computer science, for example in linguistics, forensics, and legal reasoning.

To illustrate once more an inductive definition of a BNF, we define the BNF for decimal numbers.

Example 4.11 (BNF for Decimal Numbers).

$$\text{<DecimalNumber>} ::= \text{<Digit>} \mid \text{<DecimalNumber><Digit>}$$
$$\text{<Digit>} ::= \mathbf{0} \mid \ldots \mid \mathbf{9} \tag{4.7}$$

As in the previous section, we allow in the above formalization for leading zeros in representations of decimal numbers. We leave it as an exercise to modify this BNF so that leading zeros are eliminated, similarly to how this was done for another BNF on page 140.

The BNF for Boolean terms (recall Definition 4.5) may be defined as follows.

Example 4.12 (BNF for Boolean Terms).

$$\text{<BT>} ::= \mathsf{T} \mid \mathsf{F} \mid \text{<V>} \mid \neg\text{<BT>} \mid (\text{<BT>} \wedge \text{<BT>}) \mid (\text{<BT>} \vee \text{<BT>})$$
$$\text{<V>} ::= X_0 \mid X_1 \mid \ldots$$

We note that the BNF for Boolean terms specifies an infinite set of variables $\{X_i \mid i \geq 0\}$, making use of the stylized dots notation "$\ldots$". Strictly speaking, this is not allowed in BNFs: the set of rules, the set of terminal symbols, and the set of non-terminal symbols should be finite. Note that this does not mean that the set of finite words generated by a BNF is finite; it is typically infinite and countably so. However, we can avoid this use of the dots notation "..." in the above BNF by naming variables explicitly through decimal numbers. For this, we need to adjust the BNF from Example 4.12 as follows:

$$<BT> ::= T \mid F \mid <V> \mid \neg<BT> \mid (<BT> \wedge <BT>) \mid (<BT> \vee <BT>)$$
$$<DecimalNumber> ::= <Digit> \mid <DecimalNumber><Digit>$$
$$<Digit> ::= \mathbf{0} \mid \ldots \mid \mathbf{9}$$
$$<V> ::= X<DecimalNumber>$$

In this BNF, we may now derive variable names $X\mathbf{0}$, $X\mathbf{1}$, $X\mathbf{2}$, ..., for example through the derivation sequence

$$<V> \Longrightarrow X<DecimalNumber> \Longrightarrow X<Digit> \Longrightarrow X\mathbf{0}$$

In practice, one often uses a production rule of the form

$$<V> ::= \mathbf{v} \in \mathscr{V}$$

where $\mathbf{v}$ is a so-called *metavariable* for a separately specified set of variable names, for example $\mathscr{V} = \{X_0, X_1, \ldots\}$. At the level of BNFs, such metavariables are treated in a special way: Like terminal symbols, they do not occur on the left-hand side of any rule of a BNF, and they can be instantiated, like non-terminals, but exclusively through arbitrary identifiers.

Lexical analysis and parsing The use of metavariables reflects the common approach in compiler construction, where the syntactic correctness of source code is typically examined in two steps. In the first step, the so-called *lexical analysis* divides the input, a character string, into so-called *tokens*, which we may think of as the smallest units of programming languages that carry meaning. Tokens may be single symbols such as (or {, they may be finite words – for example keywords such as if and else – or tokens may signify variables or types. Even representations of numbers, for example decimal numbers, would be seen as tokens. This first step is only concerned with recognizing and separating words in an input stream of symbols, and in that sense it is similar to the identification of words in computer linguistics. For example, it may take the input stream "234.5v67 < x4" and produce the three tokens "234.5v67", "<", and "x4". This step, however, does not *validate* whether the tokens are used according to their foreseen role. For example, it does not recognize that the token "234.5v67" is neither a valid decimal number nor a valid variable name as required for operands of "<".

The second step, called *parsing*, is concerned with validating such tokens. And this is done by appealing to BNFs. For example, it may recognize a rule saying that the above comparison with "<" needs that both "234.5v67" and "x4" are arithmetic expressions. And there may be rules saying that variable names and decimal numbers are also arithmetic expressions. So "x4" would be recognized as a valid variable name and therefore would be validated as an arithmetic expression, whereas the validation for "234.5v67", and so for the entire comparison expression, would fail.

This separation of concerns, lexical analysis and parsing, leads to much more efficient processing of input streams: the creation of tokens is a relatively lightweight task whereas parsing is complex and so becomes more manageable when confined from the level of the entire input stream down to the level of individual tokens. We refer to textbooks on *Compiler Construction* for more details on this topic [2, 49].

4.4 Inductive Schemas

Of special significance are *semantic* schemas, which assign meaning to syntactic representations that are described through inductive means, for example through an inductively defined set or a BNF. A semantic schema can then also be defined inductively, over the inductive structure of the syntactic representations to which it is intended to assign formal meaning.

Let us first consider an inductive variant of the semantic schema of Example 4.5, which we obtain by basing it on the inductive description of syntactic representations in Example 4.11.

Example 4.13 (Decimal representation of natural numbers).

- $\mathscr{R}_d =_{df} \{0,\ldots,9\}^+$
- $\mathscr{I}_d =_{df} \mathbb{N}$
- $[\![\, \cdot \,]\!]_d$ is inductively defined through
$$[\![z]\!]_d =_{df} [\![z]\!]_z$$
$$[\![wz]\!]_d =_{df} 10 \cdot [\![w]\!]_d + [\![z]\!]_d$$

It is more interesting to assign meaning to terms inductively. The elegance of this approach, which is the foundation of the so-called *denotational semantics* [74, 56], is already apparent in the semantics of Boolean terms.

Semantics of Boolean terms

The BNF for Boolean terms in Example 4.12 only prescribes the syntax and structure of Boolean terms, it is agnostic as to their meaning. We still require a *semantics* that associates formal meaning with terms defined in such a BNF. Boolean terms may contain variables, which represent atomic formulas of propositional logic. Therefore, we can assign meaning to a Boolean term only relative to given truth values for all variables that occur in it. We model such a context as a (total) function from variables to truth values. The set of all such functions over variable set $\mathcal{V}$ is therefore:

$$\mathcal{B}_{\mathcal{V}} =_{df} \{\beta \mid \beta : \mathcal{V} \to \{tt, ff\}\} \tag{4.8}$$

and we refer to such functions as *assignments* subsequently. The semantics of a Boolean term is then a function of type $\mathcal{B}_{\mathcal{V}} \to \{tt, ff\}$: for a given assignment, the Boolean term has meaning either tt or ff.

Definition 4.10 (Semantics of Boolean terms). The *semantic function* for Boolean terms is a function $[\![\,\cdot\,]\!]_B : \mathcal{BT} \to \{tt, ff\}^{\mathcal{B}_{\mathcal{V}}}$, that associates with a Boolean term, for a given assignment of truth values to variables, a truth value as its formal meaning. This is defined inductively as follows:

- $[\![T]\!]_B(\beta) =_{df} tt$
- $[\![F]\!]_B(\beta) =_{df} ff$
- $[\![X]\!]_B(\beta) =_{df} \beta(X)$ for all X in $\mathcal{V}$
- $[\![(\neg t_1)]\!]_B(\beta) =_{df} \dot{\neg}([\![t_1]\!]_B(\beta))$
- $[\![(t_1 \wedge t_2)]\!]_B(\beta) =_{df} [\![t_1]\!]_B(\beta) \mathbin{\dot{\wedge}} [\![t_2]\!]_B(\beta)$
- $[\![(t_1 \vee t_2)]\!]_B(\beta) =_{df} [\![t_1]\!]_B(\beta) \mathbin{\dot{\vee}} [\![t_2]\!]_B(\beta)$

In these definitions we use the symbols $\dot{\neg}$, $\dot{\wedge}$, $\dot{\vee}$ to denote the *semantic* logical operators. These are not to be confused with the purely syntactical logical operators "$\neg$", "$\wedge$", and "$\vee$" and are functions over the set of truth values $\{tt, ff\}$. We specify these functions through truth tables:

b_1	b_2	$\dot{\neg}b_1$	$b_1 \mathbin{\dot{\vee}} b_2$	$b_1 \mathbin{\dot{\wedge}} b_2$
ff	ff	tt	ff	ff
ff	tt	tt	tt	ff
tt	ff	ff	tt	ff
tt	tt	ff	tt	tt

□

The following example illustrates the stepwise application of this semantic function.

Example 4.14 (Application of the semantic function). Let β in $\mathscr{B}_{\mathscr{V}}$ be an assignment with $\beta(X) = f\!f$. Then we have:

$$\llbracket ((\neg X) \vee \mathsf{F}) \rrbracket_B(\beta)$$
$$= \llbracket (\neg X) \rrbracket_B(\beta) \,\dot\vee\, \llbracket \mathsf{F} \rrbracket_B(\beta) \qquad\qquad (\text{Def. } \llbracket \cdot \rrbracket_B \text{ for Disjunction})$$
$$= \dot\neg(\llbracket X \rrbracket_B(\beta)) \,\dot\vee\, \llbracket \mathsf{F} \rrbracket_B(\beta) \qquad\qquad (\text{Def. } \llbracket \cdot \rrbracket_B \text{ for Negation})$$
$$= \dot\neg\beta(X) \,\dot\vee\, \llbracket \mathsf{F} \rrbracket_B(\beta) \qquad\qquad (\text{Def. } \llbracket \cdot \rrbracket_B \text{ for Variables})$$
$$= \dot\neg\beta(X) \,\dot\vee\, f\!f \qquad\qquad (\text{Def. } \llbracket \cdot \rrbracket_B \text{ for Constants } \mathsf{F})$$
$$= \dot\neg f\!f \,\dot\vee\, f\!f \qquad\qquad\qquad (\text{Evaluation of } \beta(X))$$
$$= tt \qquad\qquad\qquad\qquad (\text{Evaluation with operators } \dot\neg, \dot\vee)$$

In Section 2.1.1, we defined the semantic equivalence of formulas of propositional logic by stipulating that the entries of the corresponding truth tables agree. But this definition based on truth tables is only of intuitive character. For example, $\mathscr{A} \vee \neg\mathscr{A}$ and $(\mathscr{B} \to \mathscr{C}) \vee (\mathscr{C} \to \mathscr{B})$ are semantically equivalent but do not even share variables for atomic propositional formulas. We can use the inductive semantic schema of Definition 4.10 to provide a formal definition of semantic equivalence.

Definition 4.11 (Semantic equivalence of Boolean terms). Let t_1 and t_2 be Boolean terms in $\mathscr{BT}$ over a variable set $\mathscr{V}$. Then terms t_1 and t_2 are *semantically equivalent* if, and only if, we have that

$$\forall \beta \in \mathscr{B}_{\mathscr{V}}.\ \llbracket t_1 \rrbracket_B(\beta) \,=\, \llbracket t_2 \rrbracket_B(\beta)$$

We write $t_1 \equiv t_2$ whenever t_1 and t_2 are semantically equivalent in this sense. $\qquad\square$

It is easy to see that this formal definition correctly captures the informal one: each assignment β in $\mathscr{B}_{\mathscr{V}}$ corresponds to a row of the corresponding truth table. But note that this requires that the variables used in the truth tables are those that occur in *either* of the two formulas. Then this also captures the equivalence of $\mathscr{A} \vee \neg\mathscr{A}$ and $(\mathscr{B} \to \mathscr{C}) \vee (\mathscr{C} \to \mathscr{B})$ correctly.

Inductive definitions are a core principle for the conceptualization of structures in Informatics, be they programming languages, logics, calculi, and so forth. The full power of this principle becomes apparent when it is used for the definition of representations, semantics, properties, or algorithmic procedures – as we have done thoroughly in this chapter. In the next chapter, we will see that this inductive approach is supported by deductive proof techniques that are tailored for proving properties of inductively defined structures. This leads to a solid foundation of a conceptually clean and elegant architecture, in which structures can be defined and proved correct inductively, so that more complex structures can then be defined and proved – again inductively. In this sense the BNF-based inductive approach can be regarded as the foundation of a domain specific framework for designing new programming languages, logics, and other such formalisms in a controllable fashion.

4.5 Reflections: Exploring Covered Topics More Deeply

4.5.1 Notations and Standards

The interplay between representation and meaning is far reaching and, at times, problematic. We can see this already when simply observing the dual or multiple use of standard notation: the same symbol may have very different interpretations in different contexts. An especially striking example of this is the two variants of the definition of prime numbers on page 43, which use the symbol "|" in a notational standard for defining sets (so-called *set comprehension*) and as a divisibility relation over integers. In fact, the symbol "|" has many other established usages. For example, it serves as left- and right-hand side of the absolute value operator and for the operator that denotes the length of finite words. Also, we used it to separate rules of BNFs that share the same left-hand side as clauses. Fortunately, the given context and – if need be – the experience of the reader will determine the intended and correct interpretation of a symbol.

Standardization alone, be it of symbols or more complex artifacts such as software, cannot always prevent conflict though. And adhering to a standard does not necessarily mean that interpretations are compatible. In software engineering practice, we often come to this painful realization when we mean to combine solutions that each are labeled as standard-compliant. For example, physical measurements have national and international standards but safety-critical systems can fail when we combine input values that each are compliant with their own standard (e.g., a metric scale in millimeters and a non-metric scale in feet and inches) but where this combination does not correctly relate such different standards. Human beings can help with identifying such misalignments but this comes at a great economic cost and does not rule out critical errors. This explains why we see an increased research and development effort for interface and adaptor languages[16] that can bridge such semantic gaps, for example in federating or merging IT infrastructures of different organizations. And it also explains why we see a push in industrial research to create so-called *scientific types* for scientific computations, where variables not only are types such as integers, floating point numbers, and so forth but where they also have an associated physical dimension that is standardized.[17]

---

[16] See http://en.wikipedia.org/wiki/Interface_description_language
[17] See http://typeregistry.org

4.5.2 Linear Lists in Functional Programming Languages

If you have already had exposure to programming and programming languages, then you are likely to be familiar with the concept of *linear lists*. A linear list is a sequence of n elements, where n is a natural number. In such a sequence each element has a fixed position, referred to as its *index*. The index of an n-element sequence ranges from 0 to $n-1$, where the first (i.e., leftmost) element has index 0 and the rightmost/last element has index $n-1$. Informally, we may write such a linear list as in $[a, aba, bba, abab]$. The latter is a list of four finite words over the alphabet $\{a, b\}$ where the finite word bba has index 2.

In imperative and object-oriented programming languages such as C, C++, Java, or C# linear lists are often implemented through arrays or so-called dynamic lists. In *functional* programming languages such as Haskell [38] one often works with an *inductive* definition of linear lists. The structure of that definition is very similar to the axioms of Peano Arithmetic studied in Section 4.1.1.

We may define linear lists inductively over a specific set of elements, for example the integers $\mathbb{Z}$, as follows:

1. The empty list is a list.

2. If z is an element in $\mathbb{Z}$ and l is a list, then the structure in which z is followed by l is also a list.

Note that this definition did not introduce concrete notation for the empty list nor for combining an element z with a list l. But in Haskell it is very easy to provide such a concrete inductive definition of linear lists over $\mathbb{Z}$: we may define a new type IntList as follows:

```
data IntList = Empty | Cons Integer IntList
```

This definition looks like a BNF with two clauses: in the first rule Empty denotes the empty list; in the second rule, the list constructor Cons has as arguments an integer (Integer) and an element of type IntList – that is to say, a linear list over integers already constructed. The informal list example $[4, 8, 15, 23, 42]$ would therefore be formally written as an element of type IntList as in

```
(Cons 4 (Cons 8 (Cons 15 (Cons 23 (Cons 42 Empty)))))
```

We may now define functions and relations that contain IntList as a type. For example, a function lstsum, which sums all the elements of a list of integers, is easily defined by a case analysis that inspects the shape of the input list:

```
lstsum :: IntList -> Integer
-- Base caseb: The sum of the empty list is 0
lstsum Empty = 0
-- Inductive case: The sum of the list is the result of adding
-- the first element to the sum of the remaining list
lstsum (Cons z l) = z + lstsum l
```

This style of definition (and programming) therefore uses *pattern matching*, where the patterns often capture a form of inductive reasoning – the topic of the subsequent chapter. It should be noted that the definition of `lstsum` above bears a striking resemblance to the inductive semantic definition of the summation symbol $\sum$ on page 124. Of course, Haskell has a built-in datatype for lists that is *polymorphic* (Greek for having more than one realization) in that we may specify the type of elements over which lists should be constructed. For example, the declaration `[Integer]` denotes the datatype of linear lists over integers, the empty list is written as `[]`, and the list constructor has the form `(z:l)`. Even more complex, inductively defined datatypes and algorithms may be defined in Haskell in very compact form. To illustrate this, we may define the datatype of Boolean terms and their logical operators $\wedge$ (`Conj`), $\vee$ (`Disj`), and $\neg$ (`Neg`) as well as the underlying variable set (`Var`) and constants (`Const`) as follows:

```
data BT = Const Bool      -- Boolean Constant
        | Var String      -- Variable Names
        | Neg BT          -- Negation
        | Conj BT BT      -- Conjunction
        | Disj BT BT      -- Disjunction
```

This definition is almost identical to the *mathematical* one given in Definition 2.2 on page 23. To illustrate the inductive definition of an algorithm in Haskell, we define the conversion of a Boolean term into a semantically equivalent one in which all disjunctions have been replaced with other logical operators. That this is possible follows from Theorem 5.4, which established the functional completeness of the operator set $\{\neg, \wedge\}$, see Section 5.5. In fact, we may think of the algorithm below as the computational content of the proof of that theorem:

```
disjelim :: BT -> BT
-- Eliminate all Disjunctions using De Morgan's Laws
disjelim (Disj a b)
    = (Neg (Conj (Neg (disjelim a)) (Neg (disjelim b)))))
-- Eliminate Disjunctions of Negations and Conjunctions
-- Recursively
disjelim (Neg a)    = (Neg (disjelim a))
disjelim (Conj a b) = (Conj (disjelim a) (disjelim b))
-- Constants and Variables remain unchanged
disjelim bt         = bt
```

Again, *pattern matching* gave us a very compact way of specifying this algorithm. Such Haskell code reminds us of concise mathematical notation, and this is one of the appeals of functional programming. The above examples should make clear that functional programming languages, and especially those that are based on *pattern matching*, are ideally suited for the recursive, respectively inductive, description of datatypes and algorithms. These languages are therefore a powerful tool for the compact description of inductive structures, their composition into more complex inductive structures, and structure-preserving computation over such inductively defined representations. It is therefore desirable to widen the user base of functional programming.

One way of achieving this is to include functional programming tools within popular object-oriented programming languages. Martin Oderski can be seen as a pioneer of such efforts. He proposed and developed the programming language Scala [59], which stands for *scalable programming language*. The scalability here stands for the *extensibility* of the language: programmers can invent new types, operators, and other syntactic and semantic concepts and add them as first-class citizens to the language itself.

This allows for the creation of *domain-specific languages* that are tailored for specific needs, for example the processing of contracts in the backoffice of a financial institution. Scala comes already with many constructs from functional programming, including *pattern matching*. Scala code gets compiled into Java byte code and so runs on the *Java Virtual Machine*, providing important compatibility with the needs of much of web-based IT, smart phones, and enterprise systems. It is worth noting that the financial service sector now has a great interest in using functional languages. The reason is that *pure* functional code contains no computational state, and so computations are more resilient against unanticipated errors such as variable aliasing; moreover, the absence of state makes it much easier to migrate computation – an attractive feature for companies that operate globally.

4.5.3 Unambiguous Representation

In this chapter, we learned that BNFs are a tool through which we may define, inductively, infinite sets of finite words that have a certain structure. And knowledge of such structure was used, for example, in the inductive definition of semantic schemas (see, e.g., Definition 4.10).

In our development of this material we did not touch upon the question of whether the same word may have more than one derivation in a given BNF. Intuitively, it seems more important to know whether there is at least one such derivation – *parsing* should be able to answer this for us. However, we will now see that it is also important to know whether a BNF allows for more than one derivation of at least one of the words it can generate. If so, this creates an *ambiguity*: there will then be more than one way of explaining why a finite word is in the inductively defined set. Although such ambiguity may seem innocuous, it turns out that it can corrupt the semantic interpretation of elements. We will now discuss BNFs for Boolean terms to demonstrate that an *ambiguity* in the derivation of syntactic objects (here Boolean terms) can lead to an *ambiguity* in the meaning of such syntactic entities. Clearly, the latter is unacceptable for Boolean terms where the meanings are the truth values tt and ff.

The BNF of Boolean terms in Example 4.12 made explicit use of parentheses. This leads to a proliferation of parentheses for larger Boolean terms that makes such terms hard to read for humans. In Section 2.1.1, we introduced precedence rules (for example, $\wedge$ has priority over $\vee$). In combination with the associativity of the logical operators $\wedge$ and $\vee$, this allows us to interpret the term $X_1 \wedge X_2 \wedge X_3 \vee X_4$ uniquely as a Boolean term, and this leads to a much simpler to read form than the one that uses explicit parentheses everywhere, as in $(((X_1 \wedge X_2) \wedge X_3) \vee X_4)$.

A first attempt at gaining a similar advantage through a BNF may be to allow the use of parentheses (to make that use optional) but to not insist on their use. A BNF that explores this idea may be the following:

$$\text{<BT'>} ::= \text{T} \mid \text{F} \mid \text{<V>} \mid \neg\text{<BT'>} \mid \text{<BT'>} \wedge \text{<BT'>} \mid$$
$$\text{<BT'>} \vee \text{<BT'>} \mid (\text{<BT'>})$$

Let us now attempt to define a formal semantics for these Boolean terms, similarly to how we did this in Definition 4.10. A simple example illustrates that the resulting semantic schema has serious issues:

$$
\begin{aligned}
[\![\text{F} \wedge \text{T} \vee \text{T}]\!]_B(\beta) &= [\![\text{F}]\!]_B(\beta) \wedge [\![\text{T} \vee \text{T}]\!]_B T(\beta) \\
&= \mathit{ff} \wedge \mathit{tt} \\
&= \mathit{ff} \\
&\neq [\![\text{F} \wedge \text{T}]\!]_B(\beta) \dot{\vee} [\![\text{T}]\!]_B(\beta) \\
&= \mathit{ff} \dot{\vee} \mathit{tt} \\
&= \mathit{tt}
\end{aligned}
$$

The semantic schema $[\![\cdot]\!]_{\text{BT'}}$ is therefore *not well defined*: it assigns to the same word more than one meaning, and these meanings are in contradiction to each other. Therefore, this is *not* a semantic *function*. The problem is not that this new BNF defines terms that cannot be seen to be Boolean terms or that it does not generate some terms that we would accept to be Boolean terms: it defines exactly the same Boolean terms as the unambiguous BNF of Example 4.12. But the above BNF does not adequately represent important *semantic* aspects such as the precedence and associativity of logical operators, and this leads to problems in the semantic interpretation of such terms.

However, these issues are surmountable. The solution is to capture the needed semantic aspects within the syntactic level of the BNF. For our above example, this means that we introduce different non-terminal symbols that correspond to the precedence orders of logical operators. With some reflections (whose articulation we leave as an exercise), we can convince ourselves that the following BNF now gives rise to a correct semantic function:

$$\text{<DBT>} ::= \text{<DBT>} \vee \text{<CBT>} \mid \text{<CBT>}$$
$$\text{<CBT>} ::= \text{<CBT>} \wedge \text{<ABT>} \mid \text{<ABT>}$$
$$\text{<ABT>} ::= \neg\text{<ABT>} \mid \text{T} \mid \text{F} \mid \text{<V>} \mid (\text{<DBT>})$$

The combination of hierarchical descent and recursive definition in this BNF is noteworthy. Most BNFs we encountered so far either contained rules that were directly recursive, as in $\text{<BT>} ::= \neg\text{<BT>}$, or they referred to non-terminal symbols of *decreasing* complexity, as in <DecimalNumber> ::= <Digit>. The enforcement of precedence rules in the above BNF, however, contains *indirect* recursion: upon leaving the level of a disjunctive Boolean term (for example, because an operator which has higher precedence than ∨ is used), it is possible to re-enter that level. Therefore, such a re-entry is only possible through the explicit insertion of parentheses, which is syntactically expressed in the BNF by the rule <ABT> ::= (<DBT>). In that sense, we may see the above BNF as a syntactic justification of the semantic precedence rules and the resulting ability to omit unnecessary parentheses.

4.5.4 Some BNFs Are Better than Others

In order to illustrate how critical good representation engineering is, let us consider two BNFs for the set $\mathbb{N} \times \mathbb{N}$.

The first BNF has an explicit operator • that allows us to transparently realize that two natural numbers are being paired:

$$<Nat2> ::= <Nat> \bullet <Nat>$$
$$<Nat> ::= <Digit> \mid <Digit><Nat>$$
$$<Digit> ::= \mathbf{0} \mid \ldots \mid \mathbf{9}$$

In the second BNF below, the role of the • operator is hidden in the internal structure of the BNF.

$$<Nat2'> ::= <Digit><Nat2'> \mid <Digit><DotNat>$$
$$<DotNat> ::= \bullet<Nat>$$
$$<Nat> ::= <Digit> \mid <Digit><Nat>$$
$$<Digit> ::= \mathbf{0} \mid \ldots \mid \mathbf{9}$$

This shows that there is certainly scope for different inductive definitions of the same set of finite words. Good representation engineering exploits such scope, similarly to tactics one may employ in the game of golf – where one wants to prepare for a good position on the fairway. Let us make this concrete. Suppose that we want to define two arithmetic functions, addition and multiplication

$$plus : \mathbb{N} \times \mathbb{N} \to \mathbb{N}, \qquad mult : \mathbb{N} \times \mathbb{N} \to \mathbb{N}$$

by appeal to either of the two above BNFs as representations of the set $\mathbb{N} \times \mathbb{N}$. For the first BNF, this is very simple (like playing a golf stroke from the fairway) and can be realized through the following pattern

$$plus(n \bullet m) =_{df} Add(n,m) \qquad mult(n \bullet m) =_{df} Mult(n,m)$$

where *Add* and *Mult* are functions already defined as the realization of the intuitive functions of addition and multiplication for natural numbers. A corresponding definition based on the second BNF above, however, is rather complicated (like having to play from a bunker or out of the rough).

In the next chapter, we will see that these problematic aspects of ambiguity and complexity extend to the corresponding domain of inductive proofs. For such proofs, it will be of great advantage to define an order for the enumeration of elements defined in a BNF. This order can then serve as a foundation for inductive correctness arguments or proofs (we refer to Sections 5.7.4 and 6.1 for further details on this). The good news is that a consistent and competent approach, ranging from the clean and elegant definition of representations and the definition of required properties to the conduct of inductive proofs, generally ensures that we stay on the metaphorical fairway when using inductive techniques.

4.6 Learning Outcomes

The study of this chapter will give you the abilities and learning outcomes discussed below.

- You will know and be able to apply the Peano Axiomatization of natural numbers:

  - Which "building blocks" (atoms and operations) are required for this axiomatization?
  - Why is there no need to specify arithmetic operations directly in the Peano Axioms?

- You will be familiar with inductive definitions of sets, and be able to use them confidently:

  - What is the schema used to define sets inductively?
  - What significance do inductively defined sets have in Informatics?

- You will understand how inductive functions and properties can be defined over inductively defined sets:

  - What is the schema for defining such functions and properties inductively?
  - What significance do inductively defined functions and properties have in Informatics?

- You will have gained a deeper understanding of the importance of Compositionality:

  - Why is compositionality so important?
  - How can we achieve compositionality?
  - What are the central application domains for compositionality?
  - What do these questions above have to do with homomorphisms?

- You will have appreciated the underlying concept of the separation of syntax (representation) and semantics (meaning):

  - Why is this separation especially important in Informatics?
  - How can we best capture semantics formally?
  - How do semantic schemata used in Informatics differ from those pertaining to real-world phenomena? And why are such differences indispensable in Informatics?

- You will be able to understand and use the Backus–Naur Form (BNF) for the inductive definition of syntactic structures:

  - What elements make up a BNF? How do we specify its rules?
  - How can we construct (terminal) words out of a non-terminal symbol?
  - How can we associate a formal semantics with syntactic structures that are defined through a BNF?

4.7 Exercises

Exercise 4.1 (Inductive Definitions in Arithmetic).
Provide an inductive definition of the exponentiation of natural numbers, that is for the computation of n^m. This definition should be based on the Peano Axioms of natural numbers (see Definition 4.1) and on the inductive definitions of addition and multiplication of natural numbers (see Definitions 4.2 and 4.3).

Exercise 4.2 (Equality of formal languages).
Prove the equality of formal languages stated in (4.2).

Exercise 4.3 (Consistent definitions of Kleene hulls).
Let A be a finite alphabet.

1. Let ε not be in the finite alphabet A. Show that the two different definitions of Kleene hulls for alphabet A, in (4.1) and in (4.3), render the same formal languages.
2. Show that the argument from the last item no longer works if ε is an element of A: i.e., show that then these two definitions produce different formal languages.

Exercise 4.4 (Injective but not surjective semantic schema).
Show that the semantic schema $[\![\cdot]\!]_{bs} : \mathcal{R}_{bs} \rightarrow \mathfrak{P}(\mathbb{N})$, discussed on page 142, is injective but not surjective.

Exercise 4.5 (Inductive definition of set of partitions).
Provide an inductive definition of the number of partitions with exactly k partition classes (see Section 3.3.1) of a set of n elements, where $1 \leq k \leq n$. Use this definition to determine the total number of partitions of a set with five elements.

Exercise 4.6 (Well-definedness of inductively defined sets).
Recall Definition 4.4 from page 131 for some sets of atomic building blocks At and operators Op, respectively.

1. Let $\mathcal{M}_1$ and $\mathcal{M}_2$ be sets that both satisfy the rules (C_1) and (C_2). Show that the intersection $\mathcal{M}_1 \cap \mathcal{M}_2$ is also a set that satisfies the rules (C_1) and (C_2).
2. Show that the argument in the previous item generalizes to infinitely many such sets: let I be a set such that, for all i in I, the set $\mathcal{M}_i$ is closed under rules (C_1) and (C_2). Then $\bigcap_{i \in I} \mathcal{M}_i$ is also a set closed under rules (C_1) and (C_2).
3. Now consider the class $\mathcal{C} = \{\mathcal{M}' \mid \mathcal{M}'$ is a set closed under rules (C_1) and $(C_2)\}$. You may assume that $\mathcal{C}$ is non-empty.

   a. Show that $\bigcap \mathcal{C}$ is well defined and a set. Hint: take some element $\mathcal{M}^*$ of $\mathcal{C}$, which is a set by definition of $\mathcal{C}$. Then show that $\bigcap \mathcal{C}$ equals $\bigcap \{\mathcal{M}' \cap \mathcal{M}^* \mid \mathcal{M}' \in \mathcal{C}\}$ and that the latter is a well-defined set.
   b. Show that $\bigcap \mathcal{C}$ satisfies the closure rules (C_1) and (C_2).
   c. Conclude that $\bigcap \mathcal{C}$ is the sought inductively defined set.

Exercise 4.7 (Parameter combinations of procedure *Hanoi*).
Recall Algorithm 2 from page 128.

1. Given that there are three rods, we saw that any one of the rods can take on the role of any other one. How many combinations of input values for the last three arguments of procedure *Hanoi* are therefore possible in principle?
2. Show that procedure *Hanoi*, when called initially with $Hanoi(n, 'S', 'A', 'T')$ will indeed reach all such possible input value combinations of its last three arguments, for sufficiently large values of n.
3. What is the minimal value of n for which all possible input combinations for the last three arguments will be reached?

Exercise 4.8 (Number of displacement steps for Algorithm 2).
Use an informal argument to show that a call $Hanoi(n, 'S', 'A', 'T')$ will generate $2^n - 1$ displacement steps.

Exercise 4.9 (Syntactic Substitution).

1. Compute the following syntactic substitutions of Boolean terms, showing the stepwise application of the inductive definition of such substitutions:

$$\Big((X \vee Y) \wedge ((\neg X) \vee Z)\Big)\Big[X \mapsto \mathsf{F}\Big]$$

2. For the resulting term from the previous item, find a semantically equivalent Boolean term that is as simple as possible.

Exercise 4.10 (Representation and its meaning).
Let us consider a set of arithmetic terms $\mathscr{AT}$ without any variables, defined through the following inductive definition:

- Every decimal number in the sense of Definition 4.5 is an (atomic) arithmetic term.
- Whenever a is an arithmetic term, then $-a$ is also an arithmetic term.
- Whenever a_1 and a_2 are arithmetic terms, then $(a_1 + a_2)$ as well as $(a_1 * a_2)$ are arithmetic terms.

Similarly to our discussion of Boolean terms, complete the following tasks for this set of arithmetic terms:

1. Specify a BNF that defines this set $\mathscr{AT}$.
2. Specify an inductive semantic schema for your BNF that maps every arithmetic term to a unique integer:
$$[\![\cdot]\!]_{AT} : \mathscr{AT} \to \mathbb{Z}$$
3. Evaluate the arithmetic term

$$(((-\mathbf{12}+\mathbf{5}) * (\mathbf{56}+-\mathbf{6})) * --\mathbf{3})$$

by using your semantic schema stepwise.

Exercise 4.11 (Assignments and evaluation of Boolean terms).
Recall the definition of assignments in (4.8). Let $\mathscr{X}$ be a subset of $\mathscr{V}$. Then we define the relation $\beta \equiv_{\mathscr{X}} \beta'$ over $\mathscr{B}_{\mathscr{V}}$ to mean that $\beta(X) = \beta'(X)$ for all X in $\mathscr{X}$. So this models that assignments agree on the set of variables $\mathscr{X}$. Show the following:

1. Relation $\equiv_{\mathscr{X}}$ is an equivalence relation over $\mathscr{B}_{\mathscr{V}}$.
2. Let t be an arbitrary Boolean term whose variables are all contained in the set $\mathscr{X}$. Then for all $\beta \equiv_{\mathscr{X}} \beta'$ we have that $[\![t]\!]_B(\beta) = [\![t]\!]_B(\beta')$. In other words, the semantic schema is invariant under $\equiv_{\mathscr{X}}$ for terms that only have variables from the set $\mathscr{X}$.

Exercise 4.12 (Characterization of formal language defined by a BNF).
Recall Example 4.10 on page 145. For the BNF in (4.6), prove that it generates the language
$$\{b^{2k} \mid k \geq 1\} \cup \{b^{2k+1}a^l \mid k \geq 0, l \geq 1\}$$

Exercise 4.13 (Modified BNF that avoids leading zeros).
Recall Example 4.11 from page 145. Redefine the BNF in (4.7) so that leading zeros are eliminated.

Chapter 5
Inductive Proofs

Induction: A tool for mastering *infinities through finite means.*

In this chapter, we want to demonstrate how inductive definitions, as discussed in Chapter 4, can be applied to prove properties of all elements from an inductively defined set, for example, to prove that all Boolean terms can be converted into an equivalent but syntactically specific form. When the property that we wish to show for all elements of an inductively defined set is also defined in terms of this inductive structure, such proofs are often quite elegant.

In this chapter, we will study several principles of inductive proof. Common to all these principles is that the proof that infinitely many objects satisfy a certain property is reduced to finitely many representative inductive inferences. Our development of this topic will not reflect the historical order in which these principles were discovered and used. Therefore, we will not start with mathematical induction over the natural numbers but begin with the most general principle of induction, *Well-Founded Induction*, or as it is also called *Noetherian Induction*, in Section 5.3. We do this since well-founded induction is both the most elementary and the most general induction principle. It does require basic concepts from order theory, though, which we will discuss in Section 5.1. The subsequent sections will then present four proof principles for induction, from the most general, Well-Founded Induction, to more specialized ones, where the most restricted form – mathematical induction over the natural numbers, which resembles Peano Axiom P5 – may be familiar to you from your school days already.

5.1 Order Relations

Orders occur in many application contexts. The sets of numbers $\mathbb{N}$, $\mathbb{Z}$, $\mathbb{Q}$, and $\mathbb{R}$ can be equipped with an order $\leq$ that allows us to compare numbers with respect to their "size". The power set of a set, as another example, can be furnished with the set inclusion relation $\subseteq$ where "larger" sets have more elements but also contain all the elements of smaller sets. In fact, orders are the subject of an entire research

© Springer International Publishing AG, part of Springer Nature 2018

B. Steffen et al., *Mathematical Foundations of Advanced Informatics*,

https://doi.org/10.1007/978-3-319-68397-3_5

area in Informatics that subsumes the important algebraic structures called *complete lattices*, which we will treat in more detail in the second volume of this trilogy.

5.1.1 Partial Orders

The order relation $\leq$ on sets of numbers, for example on $\mathbb{Q}$, is such that any two numbers are comparable: for all p and q in $\mathbb{Q}$, we have that $p \leq q$ or $q \leq p$. In contrast, the order of subset inclusion over a power set has incomparable elements as soon as the underlying set has more than one element. Let $A = \{1,2\}$: then we have neither $\{1\} \subseteq \{2\}$ nor $\{2\} \subseteq \{1\}$. We refer to orders in which not all elements may be comparable as *partial orders*, whereas orders such as $\leq$ over $\mathbb{Q}$ in which every two elements are comparable are called *total orders*. Note that a total order is also a partial order since the latter structure does not insist that there are incomparable elements. We now formalize the concept of a partial order as a binary relation with specific properties.

Definition 5.1 (Partial order). A homogeneous relation $\sqsubseteq \subseteq A \times A$ is called a *partial order* over set A if, and only if

1. $\sqsubseteq$ is *reflexive*: $\forall a \in A. \; a \sqsubseteq a$

2. $\sqsubseteq$ is *antisymmetric*: $\forall a_1, a_2 \in A. \; a_1 \sqsubseteq a_2 \wedge a_2 \sqsubseteq a_1 \Rightarrow a_1 = a_2$

3. $\sqsubseteq$ is *transitive*: $\forall a_1, a_2, a_3 \in A. \; a_1 \sqsubseteq a_2 \wedge a_2 \sqsubseteq a_3 \Rightarrow a_1 \sqsubseteq a_3$ □

Let us emphasize that the underlying set A is given prior to and independent of any binary relation $\sqsubseteq$ that may order elements of such a set A. Elements of sets, as such, are unordered. In particular, there may be more than one – indeed many – partial orders over a set A. Therefore, it is important to choose an order relation that is right for the application in mind. In practice, therefore, we are interested in a set A paired with a partial order $\sqsubseteq$. We write this as $(A, \sqsubseteq)$ and refer to this pair, by slight abuse of language, as a *partially ordered set*.

It is pretty easy to establish that $(\mathfrak{P}(A), \subseteq)$ is a partially ordered set for each underlying set A. Another example is the pair $(\mathbb{N}, |)$ where $|$ is the divisibility relation defined on page 32. We leave it as exercises to prove that these are indeed partial orders.

On the set of natural numbers $\mathbb{N}$, we all have an intuitive understanding of the total order $0 \leq 1 \leq 2 \leq ....$[1] However, this order relation is not an explicit part of the Peano Axioms (see Definition 4.1). But we may appeal to these axioms to define that total order as follows.[2]

---

[1] We will formalize total orders in Section 5.1.3. But we already said that total orders are a special case of partial orders.

[2] Note that the definition of this order rests on the definition of addition of natural numbers; for the latter recall Definition 4.2.

Definition 5.2 (Total order on $\mathbb{N}$). For n and m in $\mathbb{N}$, let the relation $\leq \subseteq \mathbb{N} \times \mathbb{N}$ be defined by

$$n \leq m \Leftrightarrow_{df} \exists k \in \mathbb{N}.\ n + k = m \qquad\qquad \Box$$

Theorem 5.1. $\leq$ *is a partial order over* $\mathbb{N}$.

Proof. The proof structure below exploits the associativity, commutativity, and right cancellation rule for addition. These properties and their informal use may be familiar to you from your school days. We will postpone their formal proof until we have established the required proof methodology in Section 5.3.

Reflexivity   Let n be in $\mathbb{N}$. For $k = 0$ we then have $n + 0 = 0 + n = n$, and therefore also $n \leq n$ by the definition of $\leq$ in terms of addition.

Antisymmetry: Let n and m be in $\mathbb{N}$ such that $n \leq m$ and $m \leq n$. Appealing to the definition of $\leq$ through addition, we infer from this that there must be natural numbers k_1 and k_2 in $\mathbb{N}$ such that

$$n + k_1 = m$$
$$m + k_2 = n$$

Therefore, we can substitute the m in the second equation above with $n + k_1$ to obtain the equation $(n + k_1) + k_2 = n$. Using associativity and commutativity of addition, this implies that $(k_1 + k_2) + n = n$. From the definition of addition over natural numbers – see Definition 4.2(a) – we get from this that

$$(k_1 + k_2) + n = 0 + n$$

Using the right cancellation rule that $x_1 + y = x_2 + y$ implies $x_1 = x_2$ for all natural numbers x_1, x_2, and y, we obtain from the above that

$$k_1 + k_2 = 0 \qquad\qquad (5.1)$$

If we can now show that k_1 equals 0, then n equals m as desired.
Proof by Contradiction: Assume that k_1 is not equal to 0. By Lemma 4.1, there is then a natural number k_1' with $k_1 = \mathfrak{s}(k_1')$. From the definition of addition over natural numbers, we then also get:

$$k_1 + k_2 = \mathfrak{s}(k_1') + k_2 \overset{\text{Def.4.2}(b)}{=} \mathfrak{s}(k_1' + k_2)$$

But then we learn that $k_1 + k_2$ is the successor of some natural number. By Peano Axiom P3, this would mean that $k_1 + k_2$ is different from 0, a contradiction to (5.1).

Transitivity: Let n, m, and p be in $\mathbb{N}$ with $n \leq m$ and $m \leq p$. Then there are natural numbers k_1 and k_2 in $\mathbb{N}$ with

$$n + k_1 = m$$
$$m + k_2 = p$$

If we substitute the m in the second equation with $n + k_1$ from the first equation, we obtain the equation $(n + k_1) + k_2 = p$. Using the associativity of addition, this implies that

$$n + (k_1 + k_2) = p$$

By appeal to the definition of $\leq$ through addition, this gives us that $n \leq p$.   □

Formal proof of Theorem 5.1 The proof of Theorem 5.1 was informal; for example, it appealed informally to properties of addition. But it was also informal in a more important, potentially problematic sense. A formal version of this theorem and its proof, covered below in Theorem 5.5, appeals to the principle of mathematical induction. Said principle is a special case of Well-Founded Induction. The latter principle assumes the existence of a special partial order that, for mathematical induction, turns out to be $\leq$ of Theorem 5.1. This creates an apparent circularity of reasoning: we need mathematical induction to prove that $\leq$ is a partial order, and we need that $\leq$ is a partial order so that we obtain and may use the Proof Principle of Mathematical Induction. Fortunately, we may avoid this circularity by appeal to the Peano Axioms, notably axiom P5 – which already gives us a Proof Principle of Mathematical Induction in axiomatic form. Therefore, we may say that the proof of Theorem 5.1 is complete, despite its informality, and that we therefore may assume subsequently that $\leq$ is a partial order over the natural numbers $\mathbb{N}$.

5.1.2 Preorders

At times, it is too restrictive to insist that orders be antisymmetric. Recall that this property stipulates that two elements must be equal if one is related to the other and vice versa. For example, we would think that the height in centimeters orders the set of people, and there is more than one person with height 172 cm – ruling out the utility of antisymmetry in this application. Similarly, the divisibility relation over integers loses antisymmetry; for example, we have both $-1|1$ and $1|-1$ although 1 and -1 are clearly different. Another motivating example is the relation $\leq$ (recall Definition 3.11) that compares the sizes of sets: it is easy to see that this is a reflexive and transitive relation; but its special case, the equipotency relation $\cong$, is also symmetric and so $\leq$ is not antisymmetric. For example, we have $\{1,2\} \cong \{a,b\}$ and $\{a,b\} \cong \{1,2\}$ although $\{a,b\}$ and $\{1,2\}$ are different sets.

If we weaken Definition 5.1 by not insisting on antisymmetry, then we arrive at the notion of a *preorder*. The validity of implications on the set of Boolean terms is a typical example of a preorder; that is to say, the relation $\underset{\sim}{\sqsubseteq}$ over Boolean terms t_1

and t_2, defined as $t_1 \precsim t_2$ holds iff $t_1 \Rightarrow t_2$ is logically valid, is a preorder. The proof that this is a preorder is left as an exercise. We note here that this relation $\precsim$ is not a partial order: syntactically distinct but semantically equivalent Boolean terms give rise to violations of antisymmetry. For example, we have $X \precsim (X \wedge (X \vee Y))$ and $(X \wedge (X \vee Y)) \precsim X$ whereas X and $(X \wedge (X \vee Y))$ are different Boolean terms.

In general, for every preorder $\precsim \subseteq A \times A$ we may define an equivalence relation

$$a_1 \sim a_2 \Leftrightarrow_{df} a_1 \precsim a_2 \wedge a_2 \precsim a_1 \tag{5.2}$$

over set A; this $\sim$ is sometimes referred to as the *kernel* of the preorder $\precsim$.

Conversely, let $\sim$ be an equivalence relation over a set A. Then every partial order $\sqsubseteq$ on the set of equivalence classes $A/\sim$ *induces* a preorder $\precsim$ on set A:

$$a_1 \precsim a_2 \Leftrightarrow_{df} [a_1]_\sim \sqsubseteq [a_2]_\sim$$

In practice, the distinction between partial order and preorder may at times not be made explicitly. This is so since we may turn any preorder $\precsim$ into a partial order over the set of equivalence classes with respect to the preorder $\precsim$. It is therefore often merely a question of the level of abstraction used to determine whether we work with a preorder or a partial order.

5.1.3 Total Orders

A preorder $\precsim \subseteq A \times A$ in which all elements are comparable, i.e., where we have

$$\forall a_1, a_2 \in A. \ a_1 \precsim a_2 \vee a_2 \precsim a_1 \tag{5.3}$$

is called a *total* preorder or a *preference order*. If the preorder $\precsim$ is also a partial order, then we call $\precsim$ a *total order* or a *linear order*. It seems a bit confusing to call partial orders that satisfy (5.3) *total* orders. We should read the term "partial" perhaps as meaning "partial or total". The alternative would be to define partial orders as those that do not satisfy (5.3), which seems even more confusing.

Above, we discussed an order based on the height of people. This is an example of a total preorder. An example of a linear order is $(\mathbb{N}, \leq)$ the ordering of the natural numbers (see Figure 5.2(a)).

5.1.4 Strict Orders

For a given preorder $\precsim$, we may define the corresponding *strict* preorder as:

$$a_1 \sqsubset a_2 \Leftrightarrow_{df} a_1 \mathrel{\underset{\sim}{\sqsubseteq}} a_2 \wedge a_1 \mathrel{\not\sim} a_2 \tag{5.4}$$

where $\sim$ is the equivalence relation defined in (5.2). The thus-defined strict preorder has the following properties characteristic of strict orders.

Lemma 5.1 (Properties of a strict order). *Let $\mathrel{\underset{\sim}{\sqsubseteq}}$ be a preorder on a non-empty set A. Then the relation $\sqsubset$ defined as in (5.4) satisfies:*

1. $\sqsubset$ *is asymmetric:* $\forall a_1, a_2 \in A. \ a_1 \sqsubset a_2 \Rightarrow a_2 \not\sqsubset a_1$

2. $\sqsubset$ *is transitive:* $\forall a_1, a_2, a_3 \in A. \ a_1 \sqsubset a_2 \wedge a_2 \sqsubset a_3 \Rightarrow a_1 \sqsubset a_3$

Proof. 1. Let a_1 and a_2 be in A with $a_1 \sqsubset a_2$. From the definition in (5.4) we immediately obtain:

$$a_1 \mathrel{\underset{\sim}{\sqsubseteq}} a_2 \wedge a_1 \mathrel{\not\sim} a_2$$

By the definition of $\sim$ in (5.2), this implies that $a_2 \mathrel{\underset{\sim}{\not\sqsubseteq}} a_1$ and therefore also $a_2 \not\sqsubset a_1$ by appeal to (5.4).

2. Let a_1, a_2, and a_3 be in A with $a_1 \sqsubset a_2$ and $a_2 \sqsubset a_3$. By (5.4), this implies that $a_1 \mathrel{\underset{\sim}{\sqsubseteq}} a_2$ and $a_2 \mathrel{\underset{\sim}{\sqsubseteq}} a_3$ hold. By transitivity of $\mathrel{\underset{\sim}{\sqsubseteq}}$ we therefore obtain $a_1 \mathrel{\underset{\sim}{\sqsubseteq}} a_3$. It remains to show that $a_1 \mathrel{\not\sim} a_3$. Proof by Contradiction: assume that $a_1 \sim a_3$. Then we also have $a_3 \mathrel{\underset{\sim}{\sqsubseteq}} a_1$. Now, since $a_1 \mathrel{\underset{\sim}{\sqsubseteq}} a_2$, we again appeal to the transitivity of $\mathrel{\underset{\sim}{\sqsubseteq}}$ to get $a_3 \mathrel{\underset{\sim}{\sqsubseteq}} a_2$. But the latter, together with $a_2 \mathrel{\underset{\sim}{\sqsubseteq}} a_3$ renders $a_2 \sim a_3$, a contradiction to the assumption that $a_2 \sqsubset a_3$. $\qquad\square$

We would like to stress that the concept of *a*symmetry should not be confused with that of *anti*symmetry. The latter is required for partial orders, which are also reflexive. The former, however, implies that an asymmetric relation is also *irreflexive*, i.e., $\forall a \in A. \ a \not\sqsubset a$. In particular, such a relation cannot be reflexive, and so also cannot be a partial order.

Above, we saw that partial orders induce a strict order – an asymmetric and transitive relation by virtue of Lemma 5.1. Conversely, a relation that is asymmetric and transitive induces a partial order as follows:[3]

$$a_1 \sqsubseteq a_2 \Leftrightarrow_{df} a_1 \sqsubset a_2 \vee a_1 = a_2 \tag{5.5}$$

We may reduce a strict order $\sqsubset$ to its pairs that contain immediate neighbors in that they cannot be interpolated with a third element in that order. We write $\sqsubset_N$ for this relation:

$$a_1 \sqsubset_N a_2 \Leftrightarrow_{df} a_1 \sqsubset a_2 \wedge \not\exists a_3 \in A. \ a_1 \sqsubset a_3 \wedge a_3 \sqsubset a_2 \tag{5.6}$$

From the definition of $\sqsubset_N$, it follows that $a_1 \sqsubset_N a_2$ and $a_2 \sqsubset_N a_3$ imply that $a_1 \not\sqsubset_N a_3$. Figure 5.1 illustrates the differences between the relations $\sqsubseteq$, $\sqsubset$, and $\sqsubset_N$ through the divisibility relation $\mid$ as $\sqsubseteq$ – restricted to the subset of natural numbers

---

[3] If we replace the condition $a_1 = a_2$ with $a_1 \sim a_2$ in (5.5) for some equivalence relation $\sim$ over set A, then (5.5) defines a preorder.

$\{1,2,3,4,6,12\}$. That is to say, $n|m$ is still defined to mean $\exists k \in \mathbb{N}.\ n \cdot k = m$ but n and m are now elements of $\{1,2,3,4,6,12\}$. In the figure, $n|m$ is represented by an arrow from n to m.

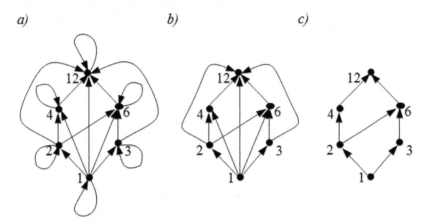

Fig. 5.1 Divisibility relation $|$ restricted to set $\{1,2,3,4,6,12\}$: a) As a partial order, b) as a strict order, and c) as neighborhood order. Arrows from n to m denote instances of divisibility $n|m$

Dense orders The definition of $\sqsubset_N$ is intuitive enough for orders over finite sets A. But when A is infinite, then $\sqsubset_N$ may be empty! Consider the "strictly less than" strict order $(\mathbb{Q}, <)$ over the rational numbers $\mathbb{Q}$. This relation has the peculiar property that for all q_1 and q_2 in $\mathbb{Q}$ with $q_1 < q_2$ there is some r in $\mathbb{Q}$ such that $q_1 < r$ and $r < q_2$. Therefore, $<_N$ is the empty relation in this case. Strict orders $\sqsubset$ for which $\sqsubset_N$ is empty are called *order dense*. They play an important role in the Theory of Computability, Topology, and other areas of Mathematics and Informatics.

For example, Cantor showed that every countable dense order that is bounded below and above is, essentially, equivalent to the dense order $(\mathbb{Q}, \leq)$ – where the notion of equivalence extends the one we studied for sets to also preserve the strict orders; we will study such homomorphisms in the second volume of this trilogy. Cantor's theorem is remarkable. For example, one consequence is that $(\mathbb{Q}, <)$ is order isomorphic to a strict subset thereof, the set of *dyadic rationals*. The latter set contains all floating-point numbers that are representable in the IEEE Standard for Floating-Point Arithmetic, and such representations need faithful mapping onto concrete hardware [45].

Our induction principles developed in this chapter will not work for dense orders. In fact, we would like to use partial orders $\sqsubseteq$ such that $\sqsubseteq$ can be recovered as the reflexive-transitive closure of $\sqsubset_N$, which requires that $\sqsubset_N$ be non-empty. Recall that, for relation $R \subseteq A \times A$, its reflexive-transitive closure is the smallest relation $R^* \subseteq A \times A$ that contains R and is reflexive and transitive (cf. Section 3.4.4).

Whenever $(\sqsubseteq_N)^*$ equals $\sqsubseteq$, then $\sqsubseteq_N$ is a much more compact representation of $\sqsubseteq$ and of great importance. The graphical representation of such a relation $\sqsubseteq_N$ is also known as a so-called *Hasse diagram* of the partial order $\sqsubseteq$. In this representation, we depict relation $\sqsubseteq_N$ through undirected links, where smaller elements are shown further down. Figure 5.2 shows the Hasse diagrams of some partial orders. Note that these diagrams allow for a unique representation of "up" and "down", due to the antisymmetry of a partial order. Reflexivity is implicit in such diagrams, as is transitivity – which "walks" from lower points along edges to upper points to get all remaining instances of the partial order that $\sqsubseteq_N$ represents.

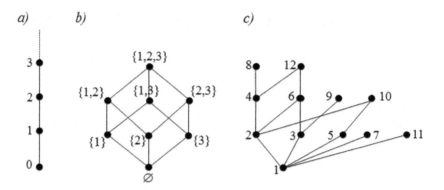

Fig. 5.2 Hasse diagrams for a) $\leq$-order over $\mathbb{N}$, b) $\subseteq$-order over $\mathfrak{P}(\{1,2,3\})$, and c) divisibility relation $|$ over $\{1,\ldots,12\}$

Hasse diagrams are therefore yet another example of the utility of separating representation (here $\sqsubseteq_N$) and meaning (the partial order $\sqsubseteq$ which equals the relation $(\sqsubseteq_N)^*$). Hasse diagrams also nicely illustrate a common phenomenon, a trade-off between the computational resources of space and time: Hasse diagrams allow us to represent a partial order in a much more compact manner, which saves a lot of storage space. But this compactness comes at the expense of time, needed for deciding or inferring whether a certain pair of elements (a,b) is in the partial order that the Hasse diagram represents.

5.2 Orders and Substructures

Let us consider subsets B of an ordered set A. Some elements of such subsets have special properties that are of interest to us. Let us first consider those elements that are at the upper and lower "boundary" of a subset B.

Definition 5.3 (Minimal, maximal elements). Let $\precsim \subseteq A \times A$ be a preorder (with strict version $\sqsubset$) and B a subset of A. An element b in B is called

1. a *minimal* element in B $\Leftrightarrow_{df}$ $\nexists b' \in B.\ b' \sqsubset b$ or

2. a *maximal* element in B $\Leftrightarrow_{df}$ $\nexists b' \in B.\ b \sqsubset b'$. □

So b in B is minimal, for example, if there is no b' different from b in B such that $b' \sqsubset b$. It should be noted that a subset B may have no, one, or more than one minimal, respectively maximal, element.

Example 5.1. For the partial order depicted in Figure 5.2(c), the subset $B = \{2,3,4,6\}$ has 2 and 3 as minimal elements, and 4 and 6 as maximal elements, respectively.

The concepts of maximal and minimal elements of a subset, for a given preorder, have a stronger version, captured in the following definition.

Definition 5.4 (Least, greatest element). Let $\sqsubseteq \subseteq A \times A$ be a preorder and B a subset of A. An element b in B is called a

1. *least* element in B $\Leftrightarrow_{df}$ $\forall b' \in B.\ b \precsim b'$ or

2. *greatest* element in B $\Leftrightarrow_{df}$ $\forall b' \in B.\ b' \precsim b$. □

It should be apparent that a least element of B is also a minimal element of B; similarly, a greatest element of B is also a maximal element of B. If $\sqsubseteq$ is a total order, so that every two elements are comparable by $\sqsubseteq$, then minimal elements are also least elements and maximal elements are also greatest elements. Using the antisymmetry of partial orders, we can also see that least and greatest elements are unique whenever they exist; therefore, we are entitled to speak of *the* least element and *the* greatest element of a partial order whenever they do exist.

Example 5.2. Consider the partial order in Figure 5.2(c). Its entire set of elements, $\{1,\ldots,12\}$, has 7, 8, 9, 10, 11, and 12 as maximal elements; in particular, there is no greatest element here. In contrast, 1 is both minimal and the least element of that partial order.

5.3 Well-Founded Induction

The Principle of *Well-Founded Induction* is the universal method for proving properties of all elements of arbitrary sets, provided that those sets are ordered in a suitable manner. This principle, which is also known under the name of *Noetherian Induction*, has fairly mild requirements on the underlying order to ensure its applicability: it merely requires a preorder in which, intuitively, we cannot make elements smaller

infinitely often. Such preorders are called *well-founded* preorders, and also are referred to as *Noetherian* preorders – in honor of the female German mathematician
Emmy Noether.[4]

Definition 5.5 (Well-founded preorder).   A preorder $\sqsubseteq\; \subseteq A \times A$ is *well-founded*
if, and only if, every non-empty subset of A has a minimal element with respect to
$\sqsubseteq$. □

The essence of the above definition is that well-founded preorders $\sqsubseteq$ rule out the
existence of infinite descending chains in set A with respect to the strict version $\sqsubset$
of $\sqsubseteq$. This *well-foundedness* is captured in the next Theorem, where we write $\sqsupset$ for
the inverse relation of $\sqsubset$.

Theorem 5.2 (Descending Chain Condition). *A preorder* $(A, \sqsubseteq)$ *is well-founded
if, and only if, there is no infinite, strictly descending chain* $x_0 \sqsupset x_1 \sqsupset x_2 \ldots$ *in A.*

Proof. We prove the equivalent, contrapositive version of this theorem, that the preorder is *not* well-founded if, and only if, there is such an infinitely descending chain:

"⇒" : Let $x_0 \sqsupset x_1 \sqsupset x_2 \ldots$ be an infinite, strictly descending chain in A. Then
$B =_{df} \{x_0, x_1, x_2 \ldots\}$ is not empty. Proof by Contradiction: assume that there is
a minimal element a_{min} in B. Then there must be an index i with $x_i = a_{min}$. Since
$x_i \sqsupset x_{i+1}$, we then get a contradiction to the fact that x_i is a minimal element
of B. Therefore, set B does not have a minimal element and so $(A, \sqsubseteq)$ is not
well-founded.

"⇐" : Let B be a non-empty subset of A such that B has no minimal element with
respect to $\sqsubseteq$. Since B is non-empty, there exists an element a in B and we set
$x_0 =_{df} a$. We may think of x_0 as a (minimal) chain that we can extend by the
successive addition of strictly descending elements. It therefore suffices to show
that such concatenations can be achieved forever. So suppose that this process
has already constructed a strictly descending chain of elements of B, the chain
$x_0 \sqsupset x_1 \ldots \sqsupset x_n$ where x_0 is the above a. Since x_n is in B and since B does not
have any minimal elements, we infer that there must be some x_{n+1} in B such
that $x_n \sqsupset x_{n+1}$; note that this argument is also valid for the initial case when n
equals 0. Therefore, we may repeat this argument infinitely often to construct an
infinite, strictly descending chain with respect to $\sqsubseteq$. (The construction is such
that all elements of the chain are in B, but this is also needed as an *invariant* for
constructing the chain; it is not needed at the global level of the proof's structure.)
□

We speak of a well-founded partial order if the partial order is well-founded when
interpreted as a preorder; recall that every partial order is a preorder as well. In particular, the characterization of well-founded preorders given in Theorem 5.2 extends
to well-founded partial orders. Here are some examples of well-founded partial orders.

---

[4] See http://en.wikipedia.org/wiki/Emmy_Noether

Example 5.3 (Well-Founded Partial Orders).

1. The partial order $\leq$ on the set of natural numbers $\mathbb{N}$ is total and well-founded. In particular, all non-empty subsets of $\mathbb{N}$ have a *least* element.

2. The subword relationship on the set of finite words A^* over alphabet A is a well-founded partial order. In every strictly descending chain, the words become shorter with each step on the chain.

3. The power set of a finite set A, ordered by set inclusion $\subseteq$, is a well-founded partial order. For each set B on a strictly descending chain, its successor on the chain has at least one element fewer than B itself.

Next, we list some partial orders that violate well-foundedness in typical ways.

Example 5.4 (Not Well-Founded Partial Orders).

1. The total partial order $\leq$ on the set $\mathbb{Z}$ of integers is not well-founded: the non-empty subset $\mathbb{Z}$ has no minimal element; and this is also the case for any infinite set of negative integers.

2. The total partial order $\leq$ on the set of non-negative rational numbers $\mathbb{Q}_{\geq 0}$ is not well-founded: the non-empty subset $\{\frac{1}{2}, \frac{1}{3}, \frac{1}{4}, \ldots\}$ has no minimal element.

3. The power set $\mathfrak{P}(\mathbb{N})$ of the infinite set $\mathbb{N}$, ordered by subset inclusion $\subseteq$, is not well-founded: the non-empty subset $\{\mathbb{N}, \mathbb{N}\backslash\{0\}, \mathbb{N}\backslash\{0,1\}, \mathbb{N}\backslash\{0,1,2\}, \ldots\}$ has no minimal element.

As already alluded to, well-founded preorders are the foundation for a fundamental proof principle, that of *Well-Founded Induction*.

Proof Principle 9 (Principle of Well-Founded Induction)

Let $\underset{\sim}{\sqsubseteq} \subseteq M \times M$ be a well-founded preorder. Let $\mathscr{A}$ be a proposition about elements of M. Suppose we can prove, for each m in M, that $\mathscr{A}$ is true for m provided $\mathscr{A}$ is true for all elements smaller than m. Then $\mathscr{A}$ is true for all m in M.

$$\Big(\forall m \in M. \underbrace{\big(\underbrace{\forall m' \in M. m' \sqsubset m \Rightarrow \mathscr{A}(m')}_{(IH)}\big) \Rightarrow \mathscr{A}(m)}_{(IS)}\Big) \Rightarrow \underbrace{\forall m \in M. \mathscr{A}(m)}_{(IC)}$$

Before we convince ourselves of the validity of this proof principle, let us make some remarks on its structure. This structure will also apply to special instances of this inductive proof principle, although such instances may present such structure in slightly different ways.

The general aim in inductive proofs is to formally prove that some property, the *Induction Claim* (IC), is universally true over some structure or set. A direct use of Proof Principle 5, which eliminates universal quantifiers, does not seem to realize the aim of proving that Induction Claim. For example, the Induction Claim in Exercise 5.14 is that each natural number $n \geq 8$ is the sum of zero or more copies of 3 and 5; and a direct way of showing this – by taking an arbitrary such number n – won't succeed.

The inductive approach transforms the Induction Claim (which we wish to prove) into a *relative* proof obligation, the so-called *Induction Step* (IS). This Induction Step is again a universally quantified proposition, which we may want to prove according to Proof Principle 5. However, the proof obligation for an arbitrarily chosen element may appeal to the so-called *Induction Hypothesis* (IH). The Induction Hypothesis states that the property we wish to show for that arbitrary element is true for all elements strictly smaller than that element. We will see below that the additional knowledge expressed in an Induction Hypothesis leads to elegant proofs, especially for inductively defined properties, sets, and functions.

In invoking this proof principle of Well-Founded Induction, the minimal elements of a structure take on a special role: a minimal element m is such that there is no m' in M with $m' \sqsubset m$. Then the Induction Hypothesis does not represent any usable knowledge, so that we then have to prove, by whatever means, that m satisfies the Induction Claim. Proofs of the Induction Claim for minimal elements are often called *Induction Base Cases*.

Let us now see why the proof principle of Well-Founded Induction, Proof Principle 9, is valid.

Theorem 5.3. *Proof Principle 9 of Well-Founded Induction is valid.*

Proof. We show the validity of this proof principle by using the Proof Principle of Contraposition. So let us assume that the Induction Claim (IC), the proposition $\forall m \in M. \ \mathscr{A}(m)$, is not true. Then we need to show that the Induction Step (IS) is also not true.

Consider the set C of all "counter examples" to that Induction Claim, defined as $C =_{df} \{c \in M \mid \neg \mathscr{A}(c)\}$. Since the Induction Claim is not true, we know that C is a non-empty subset of M. Since $\underset{\sim}{\sqsubseteq}$ is well-founded, we therefore conclude that C has some minimal element c_{min}, which we may think of as a minimal counter example to the truth of the Induction Claim $\forall m \in M. \ \mathscr{A}(m)$. Because of the minimality of c_{min}, we have that for all m' in M the relation $m' \sqsubset c_{min}$ implies that $A(m')$ is true. But this then gives us that the Induction Step

$$\left(\forall m \in M. \ \left(\forall m' \in M. \ m' \sqsubset m \Rightarrow \mathscr{A}(m') \right) \Rightarrow \mathscr{A}(m) \right) \qquad (IS)$$

is violated (as desired): the universally quantified proposition $\forall m' \in M.\ m' \sqsubset c_{min} \Rightarrow \mathscr{A}(m')$ is true (since c_{min} is a minimal counter example for $\mathscr{A}$ in M), but $\mathscr{A}(c_{min})$ is false (since c_{min} *is* a counter example for $\mathscr{A}$ in M).   □

In a first application of Well-Founded Induction, we want to prove the commutativity of addition over the natural numbers, where addition is formally defined in Definition 4.2. The Induction Claim (IC) is therefore

$$\forall n, m \in \mathbb{N}.\ n + m = m + n \tag{5.7}$$

We will summarize this property and other pertinent ones of the natural numbers further below in Theorem 5.5. The proof of that theorem also provides the standard inductive proof of the Induction Claim in (5.7) – which is an example of two nested applications of the Proof Principle of Mathematical Induction, which we discuss in detail below. However, the proof of (5.7) through Well-Founded Induction is more elegant. It merely requires a well-founded preorder on the set of pairs of natural numbers, $\mathbb{N} \times \mathbb{N}$, defined as

$$(n, m) \leq (n', m') \Leftrightarrow_{df} n \leq n' \wedge m \leq m' \tag{5.8}$$

The proof of the commutativity of addition is then as follows.

Proof. Let (n, m) be in $\mathbb{N} \times \mathbb{N}$. We use the Induction Hypothesis (IH), that all pairs (n', m') with $(n', m') < (n, m)$ already satisfy that $n' + m' = m' + n'$. Note that $<$ here refers to the strict version of $\leq$ defined in (5.8). We then consider cases that reflect the inductive structure of natural numbers:

Case 1:   Let n be 0. Then we reason that

$$n + m = 0 + m \overset{\text{(Def. 4.2.a)}}{=} m \overset{(*)}{=} m + 0 = m + n$$

The equation $(*)$ is justified since 0 is the identity element for addition over the natural numbers (cf. Theorem 5.5), a property that we here assume we can appeal to (i.e., that this has already been inductively proven).

Case 2:   Let $n \neq 0$ and $m = 0$. Then we reason similarly to Case 1 (omitted, left as an exercise).

Case 3:   let $n \neq 0$ and $m \neq 0$. Then there are n' and m' with $n = \mathsf{s}(n')$ and $m = \mathsf{s}(m')$. Then we reason that

$$
\begin{aligned}
n + m &= \mathsf{s}(n') + m \overset{\text{(Def. 4.2.b)}}{=} \mathsf{s}(n' + m) \overset{\text{(IH)}}{=} \mathsf{s}(m + n') \\
&= \mathsf{s}(\mathsf{s}(m') + n') \overset{\text{(Def. 4.2.b)}}{=} \mathsf{s}(\mathsf{s}(m' + n')) \overset{\text{(IH)}}{=} \mathsf{s}(\mathsf{s}(n' + m')) \\
&\overset{\text{(Def. 4.2.b)}}{=} \mathsf{s}(\mathsf{s}(n') + m') = \mathsf{s}(n + m') \overset{\text{(IH)}}{=} \mathsf{s}(m' + n) \\
&\overset{\text{(Def. 4.2.b)}}{=} \mathsf{s}(m') + n = m + n
\end{aligned}
$$

□

The proof of Theorem 5.3 above used a technique that combined the properties of a well-founded preorder with the Proof Principle of Proof by Contradiction. This combination deserves a Proof Principle in its own right:

Proof Principle 10 (Principle of Minimal Counter Example)

Let $\sqsubseteq \, \subseteq \, M \times M$ be a well-founded preorder. Let $\mathscr{A}$ be a proposition about elements of M. We prove that $\mathscr{A}(m)$ holds for all $m \in M$ by assuming that there are counter examples, i.e.,

$$C =_{df} \{c \in M \mid \neg \mathscr{A}(c)\} \neq \emptyset$$

$\mathscr{A}$ is proved by showing that the existence of a minimal counter example $c_0 \in C$ leads to a contradiction.

We leave the proof of the validity of this proof principle as an exercise.

In Informatics, Well-Founded Induction has a special place as a means of proving the *termination* of programs or algorithms; there is rarely an alternative to using Well-Founded Induction for termination proofs.[5] Well-Founded Induction can be used to prove that a program completes its task, and so terminates, after finitely many steps. If we think of computation steps as links in a strictly descending chain, then we will understand how Well-Founded Induction may be used for termination proofs. The next section presents three such illustrative examples of termination proofs; these examples also suggest that Well-Founded Induction is perhaps more powerful than its special case of Mathematical Induction (which we will introduce below).

5.3.1 Termination Proofs

To simplify the exposition of termination proofs, we will not define and use a concrete programming language, but work with recursive definitions (i.e., specifications) of algorithms. We then want to prove that each definition can only make finitely many recursive calls before a concrete result is computed. The fundamental idea here is to use a well-founded preorder on the arguments of such recursive definitions such that each recursive call of a definition will be made on arguments that are strictly below those for which the definition is being computed. We may then appeal to the Descending Chain Condition (recall Theorem 5.2) to argue that all recursive calls to definitions must terminate eventually.

---

[5] More sophisticated approaches can generalize this proof principle; see for example the approach introduced in [15].

First, let us consider the recursive definition of the greatest common divisor, which we may think of as a specification for an algorithm that computes this greatest common divisor for its arguments:

$$gcd : \mathbb{N} \times \mathbb{N} \to \mathbb{N}$$

$$gcd(n,m) = \begin{cases} n+m & \text{if } n=0 \text{ or } m=0 \\ gcd(n-m,m) & \text{if } 0<m\leq n \\ gcd(n,m-n) & \text{if } 0<n<m \end{cases}$$

An important thing to note is that the above definition is *complete*: each pair (n,m) of $\mathbb{N} \times \mathbb{N}$ satisfies one, and only one, of the above definitional cases. This implicitly relies on the fact that $<$ is a total order over $\mathbb{N}$. It is quite easy to see that the relation $\precsim_{sum}$ defined as

$$(n,m) \precsim_{sum} (n',m') \Leftrightarrow_{df} n+m \leq n'+m' \tag{5.9}$$

is a well-founded partial preorder. This is not a partial order since antisymmetry is violated: for example, we have $(1,3) \precsim_{sum} (2,2)$ and $(2,2) \precsim_{sum} (1,3)$ although $(1,3)$ is not equal to $(2,2)$ in $\mathbb{N} \times \mathbb{N}$. But this is not an issue for the termination proof, as we merely have to ensure that recursive calls are for arguments that strictly decrease with respect to $\precsim_{sum}$.

In the first case of this recursive definition, when n or m is 0, there is no recursive call and so the definition terminates. In the second case, we have $0<m\leq n$. The argument of the definition is the pair (n,m) and the argument of the recursive call is $(n-m,m)$. Therefore, we need to show that $(n-m,m) \precsim_{sum} (n,m)$, i.e., that $(n-m)+m<n+m$, i.e., that $n<n+m$. But this is true since $0<m$ in this case. The reasoning for the third case is similar and so omitted here. Since we showed that all cases either terminate directly (the first case) or call the definition recursively on strictly smaller arguments with respect to $\precsim_{sum}$, we infer that this inductive definition of the greatest common divisor terminates, by appeal to the Descending Chain Condition (Theorem 5.2).

A more ambitious example is the following recursive definition of an algorithm, commonly known as the *Ackermann function*:[6]

$$ack : \mathbb{N} \times \mathbb{N} \to \mathbb{N}$$

$$ack(n,m) = \begin{cases} m+1 & \text{if } n=0 \\ ack(n-1,1) & \text{if } n>0, m=0 \\ ack(n-1,ack(n,m-1)) & \text{if } n>0,\ m>0 \end{cases}$$

A well-founded total preorder suitable for proving the termination of the Ackermann function is the *lexicographical* one, where $<$ and $\leq$ refer to the usual orders over $\mathbb{N}$:

---

[6] Detailed information regarding the history and importance of the Ackermann function is found at https://en.wikipedia.org/wiki/Ackermann_function.

$$(n,m) \leq_{lex} (n',m') \Leftrightarrow_{df} n < n' \lor (n = n' \land m \leq m') \tag{5.10}$$

(We leave it as an exercise to show that this is a well-founded preorder.) First, note that the three cases of this recursive definition are mutually exclusive and complete: each pair (n,m) in $\mathbb{N} \times \mathbb{N}$ matches a unique such case. Moreover, only the last two cases contain recursion, so it suffices to consider those. In the case when $n > 0$ and $m = 0$, the recursive call is for argument $(n-1,1)$ and $(n-1,1) \leq_{lex} (n,m)$ holds by (5.10) since $n-1 < n$. In the case when $n > 0$ and $m > 0$, there are *two* recursive calls: an outermost one with argument $(n-1, ack(n, m-1))$ and an innermost one with argument $(n, m-1)$. For the latter, we have $(n, m-1) <_{lex} (n,m)$ by (5.10) since $n = n$ and $m-1 < m$ both hold. For the former, we have that $(n-1, ack(n, m-1)) <_{lex} (n,m)$ holds by (5.10) since $n-1 < n$. What is perhaps remarkable about this is that we do not know nor need to know how large the natural number $ack(n, m-1)$ is in this case (in fact, it will be very large in general). All we need is that this *is* a natural number (which was proved in the former case through a termination argument). Then we can appeal to the definition of $\leq_{lex}$ in (5.10). Therefore, we can conclude that the Ackermann function terminates for all its inputs, appealing again to the Descending Chain Condition (Theorem 5.2).

That a recursive definition always terminates does not always mean that we manage to find a well-founded preorder that can witness this fact. The two examples above may have suggested that it is always relatively easy to find such well-founded preorders. But in fact, deciding whether a given inductive definition terminates is *undecidable* in general: there is no automated, mechanical procedure that can decide this question correctly for all problem instances. Here is a prominent and concrete recursive definition for which there is no known termination proof at the time of writing, the so-called *Collatz function*.[7][8]

$$col : \mathbb{N}_{>0} \to \{1\}$$
$$col(n) = \begin{cases} 1 & \text{if } n = 1 \\ col(n/2) & \text{if } n \text{ even} \\ col(3n+1) & \text{if } n \text{ odd} \end{cases}$$

With the aid of computers, it has been verified that this definition terminates for all inputs n up to about $5 \cdot 10^{18}$. There are also termination proofs that require probabilistic assumptions about the behavior of this function, and where these probabilistic assumptions are not literally but only approximately true. But at the time of writing there is neither a termination proof nor evidence for which input values this function may not terminate on.[9]

---

[7] See https://en.wikipedia.org/wiki/Collatz_conjecture

[8] Note that this recursive definition is in accordance with the iterative version in Algorithm 1.

[9] In particular, it is presently unclear whether the relation induced by this recursive definition – where $n \sqsubset m$ if $n = col(m)$ – is well-founded or a partial order. It may well be similar to the typical

Partial vs. total Correctness In the jargon of program verification, the term *partial correctness* refers to the functional correctness of a program, for example its expected input/output behavior, however, only restricted to terminating runs. As discussed already in the introduction, the output of the Collatz function can only be the value 1 for all of its inputs. This is therefore an example of a trivial partial correctness problem: the correctness (that the output be 1) is most simple to show, assuming that the function terminates.

In order to capture the functionality of a program completely, we also have to determine whether the program does terminate for all relevant input values. The term *total correctness* expresses that a program is partially correct but also terminates on all relevant inputs.

A more precise formulation of the terms partial and total correctness benefits from the consideration of so-called *Hoare triples* of the form

$$\{pre\} \; S \; \{post\} \qquad (5.11)$$

In the above, *pre* and *post* are formulas of a suitable logic (e.g. first-order logic) that contain some non-quantified variables that may refer to identifiers in the program S. These formulas can be evaluated in computational states of the program S. We then say that the Hoare triple in (5.11) is

- *partially correct*, when the following is true: whenever program S starts in a state that satisfies *pre*, and the corresponding execution of S terminates, then formula *post* is satisfied in the terminating state.

- *totally correct*, when the following is true: whenever the program S starts in a state that satisfies *pre*, then this execution of S terminates *and* the terminating state of that execution satisfies formula *post*.

Hoare was not the only one who proposed such triples. The approach inherent to such triples can be seen as the key to *program verification* and to many techniques in *formal methods*, areas of Informatics in which one seeks to guarantee the correctness, reliability, security, or other pertinent aspects of computer systems with mathematical rigor.

5.4 Course of Values Induction

A special case of Well-Founded Induction is when we deal with the set of natural numbers $\mathbb{N}$ in their usual total order $\leq$ (and its strict version $<$) – recall Definition 5.2. In that instance, Well-Founded Induction is also called *Course of Values In-*

---

staircase pictures of Maurits Cornelis Escher [33], which suggest discrete *altitude levels* although they are cyclic, i.e., antisymmetric.

duction, Complete Induction, or *Strong Induction.* For this given total well-founded partial order, we therefore obtain a specialized proof principle:

Proof Principle 11 (Principle of Course of Values Induction)

Let $\mathscr{A}$ be a proposition about natural numbers. If the truth of $\mathscr{A}$ for a natural number is derivable from the truth of $\mathscr{A}$ for all natural numbers smaller than that number, then $\mathscr{A}$ is true for all natural numbers. Formally:

$$\Big(\forall n \in \mathbb{N}. \, (\forall m \in \mathbb{N}. \, m < n \Rightarrow \mathscr{A}(m)) \Rightarrow \mathscr{A}(n) \Big) \Rightarrow \forall n \in \mathbb{N}. \, \mathscr{A}(n)$$

The Principle of Course of Values Induction allows for a general Induction Hypothesis: to show that a property is true for a natural number $n > 0$, we may assume that this property is already true for *all* natural numbers less than n. This is in contrast to the popular Principle of *Mathematical Induction*, which we will study further below and for which we may assume only that the property is true for the predecessor $n - 1$ of n. The advantage of such a more general Induction Hypothesis becomes apparent when proving properties of the Fibonacci numbers[10].

Definition 5.6 (Fibonacci numbers).

$fib : \mathbb{N} \to \mathbb{N}$

$$fib(n) \;=\; \begin{cases} 0 & \text{if } n = 0 \\ 1 & \text{if } n = 1 \\ fib(n-2) + fib(n-1) & \text{if } n \geq 2 \end{cases} \qquad\qquad \square$$

For example, the Induction Claim

$$\forall n \in \mathbb{N}. \, fib(n) < 2^n \tag{5.12}$$

has a rather elegant and intuitive proof using the Principle of Course of Values Induction.

Proof (of Equation (5.12)). Let n be in $\mathbb{N}$. By Induction Hypothesis, we may assume that $fib(m) < 2^m$ holds for all $m < n$. We now need to consider the three definitional cases for $fib(n)$:

- Let n be 0. Then we have $fib(0) \overset{\text{(Def.)}}{=} 0 < 1 = 2^0$.

- Let n be 1. Then we get $fib(1) \overset{\text{(Def.)}}{=} 1 < 2 = 2^1$.

---

[10] See for example https://en.wikipedia.org/wiki/Fibonacci_number.

- Let $n \geq 2$. Then we reason as follows:

$$fib(n) \overset{\text{(Def.)}}{=} fib(n-2) + fib(n-1) \overset{\text{(IH)}}{<} 2^{n-2} + 2^{n-1} \leq 2^{n-1} + 2^{n-1}$$
$$= 2 \cdot 2^{n-1} = 2^n$$

Note that here we used the Induction Hypothesis twice, once for $m = n - 2$ and once for $m = n - 1$.   □

The proof of (5.12) through Mathematical Induction is rather cumbersome, and is left as an exercise. Course of Values Induction, as defined in Proof Principle 11, uses the well-founded total ordering over the natural numbers $\mathbb{N}$. Since $\mathbb{N}$ has only one minimal element, the least element 0, we have to prove $\mathscr{A}(0)$ directly without any induction hypothesis. In the Principle of Mathematical Induction discussed below, this is made explicit by considering a so-called Induction Base Case for $n = 0$. Moreover, in the above proof of (5.12) through Course of Values Induction $\mathscr{A}(1)$ does also not profit from the induction hypothesis and has to be proved separately.

The Course of Values Induction is also more elegant when proving Induction Claims for restricted ranges of $\mathbb{N}$, for example that all numbers ≥ 8 are linear combinations of 3 and 5 over $\mathbb{N}$: for all $n \geq 8$, there are a and b in $\mathbb{N}$ such that $n = a \cdot 3 + b \cdot 5$. In this case, $\mathscr{A}(n)$ is vacuously true whenever $n < 8$, as no claim is made for such numbers.

5.5 Structural Induction

We now study a special instance of Well-Founded Induction that is tailored for inductive proofs over sets that are defined inductively as in Definition 4.4. Recall that elements of such inductively defined sets M are either atoms from a specified set of atoms $\mathcal{A}t$ or are obtained from the application of operators o, from a set of operators Op, to elements of set M. Non-atomic elements obtained by such applications have the general form $o(m_1, \ldots, m_k)$, where o is in Op and takes k arguments and m_i is in M for all $1 \leq i \leq k$.

We can now prove a proposition $\mathscr{A}$ for *all* elements m in M as follows:

1. First, we prove that proposition $\mathscr{A}$ is true for all atoms a in $\mathcal{A}t$. This therefore considers each a in $\mathcal{A}t$ as an *Induction Base Case*.

2. Second, we prove for all operators o in Op that o preserves the *Invariant* of satisfying property $\mathscr{A}$. This is referred to as the *Induction Step*.

We may summarize this approach in the following proof principle:

Proof Principle 12 (Principle of Structural Induction)

Let M be a set that is inductively defined over a set of atoms $\mathcal{A}t$ and a set of operators Op. Suppose that we can prove that a property $\mathscr{A}$ is true for all atoms a in $\mathcal{A}t$; and that we can prove for all o in Op and all $m_1, m_2, \ldots, m_k$ in M that the truth of $\mathscr{A}$ for all m_i with $1 \leq i \leq k$ implies the truth of $\mathscr{A}$ for element $o(m_1, m_2, \ldots, m_k)$ in M. Then, we may infer that property $\mathscr{A}$ is true for all elements of M. Formally,

$$\left(\left((\forall a \in \mathcal{A}t. \ \mathscr{A}(a)) \wedge \right. \right.$$

$$\left. \forall o \in Op, m_1, \ldots, m_k \in M. \ \left(\mathscr{A}(m_1) \wedge \cdots \wedge \mathscr{A}(m_k) \right) \Rightarrow \mathscr{A} \left(o(m_1, \ldots, m_k) \right) \right)$$

$$\Rightarrow \forall m \in M. \ \mathscr{A}(m)$$

Structural Induction takes on a central role for inductive proofs in Informatics, as the latter often works with inductively defined sets. Structural Induction is indeed a special instance of Well-Founded Induction. To see this, consider an inductively defined set M over set of atoms $\mathcal{A}t$ and operators Op. Then we may define a relation $\sqsubset_S$ where $m \sqsubset_S m'$ captures that m' was built by some operator with m as one of its arguments. Intuitively, we think of elements as structures themselves and m is then a substructure of the structure m'. Formally, we define

$$m \sqsubset_S m' \ \Leftrightarrow_{df} \ \exists o \in Op. \ \exists m'_1. \ \ldots \exists m'_k. \ m' = o(m'_1, \ldots, m'_k) \ \wedge \ m \in \{m'_1, \ldots, m'_k\}$$

$$(5.13)$$

Let $\sqsubseteq$ be $\sqsubset_S^*$, the reflexive, transitive closure (or hull) of $\sqsubset_S$ (cf. Section 3.4.4). By construction, $\sqsubseteq$ is reflexive and transitive. But it is also anti-symmetric. We won't formally prove this here, but we note that $m \sqsubset_S m'$ means that m is somehow smaller than m'; we will formalize this later in this trilogy when we study trees. Therefore, $\sqsubseteq$ is a partial order that specifies which elements are substructures of which other elements. Since all elements of M are constructed from atoms and a finite number of applications of operators, we conclude that $\sqsubseteq$ is also well-founded. We leave it as an exercise to see that the above Proof Principle of Structural Induction is essentially the same as Well-Founded Induction for this well-founded partial order $\sqsubseteq$ on M.

The characteristic feature of Structural Induction is that it allows proofs to proceed along (not necessarily binary) *tree structures* (see also the discussions in Sections 4.4 and 5.7). This is one reason for the significance of that proof principle: tree structures are ubiquitous in Informatics. For example, they occur as abstract syntax trees [2], structural or derivation trees [2, 35], AVL-trees [17], decision trees [17], computation trees [35], and so forth.

Structural Induction is, in its essence, a rather simple proof principle to use. For the proposition $\mathscr{A}$ that we wish to prove for all elements of an inductively defined set M, we merely have to conduct a *case analysis* that first considers each atom as a case and then each operator as another case. For atoms, we have to prove $\mathscr{A}$ without any additional assumptions. For each operator op, its case considers an arbitrary element m of form $m = o(m_1, \ldots, m_k)$, where we may assume that all the k elements m_i already satisfy proposition $\mathscr{A}$. For that case, it then remains to show that m satisfies that property as well.

Functional Completeness of $\{\neg, \wedge\}$. We are now in a position to formally prove this completeness result using Structural Induction over *formulas of propositional logic* specified in Definition 2.2. The set of these formulas can be seen as an inductively defined set: atomic propositions a, b, c, and so forth make up the set of atoms. The set of operators consists of a unary operator $\neg$ (for logical negation), and binary operators $\wedge$, $\vee$, $\Rightarrow$, and $\Leftrightarrow$ with their usual logical interpretation. Subsequently, we will use the convenient infix notation for binary operators. For example, we will write $\mathscr{A} \wedge \mathscr{B}$ instead of $\wedge(\mathscr{A}, \mathscr{B})$. As remarked in Section 2.1.1 already, the set of logical operators $\{\neg, \wedge\}$ is *functionally complete*, i.e., we have the following.

Theorem 5.4. *Each formula of propositional logic ϕ is semantically equivalent to some formula ϕ', where ϕ' contains only the logical operators $\neg$ and $\wedge$.*

We now want to prove this theorem by Structural Induction, making use of the De Morgan's Laws in the process.

Proof. Let ϕ be a formula of propositional logic. We conduct a case analysis over the inductive structure of formula ϕ:

Case 1: Let ϕ be an atomic proposition a. Then ϕ does not contain any logical operators, and so we may set ϕ' to be equal to ϕ.

Case 2: Let ϕ be obtained by the logical operator $\neg$. So ϕ is equal to $\neg\psi$ for some formula ψ. By the Induction Hypothesis (IH) for ψ, we know that there exists a formula of propositional logic ψ' such that ψ and ψ' are semantically equivalent, and all logical operators in ψ' (which may have none) are from the set $\{\neg, \wedge\}$. Let us now set ϕ' to be $\neg\psi'$. Then it is easy to show that ϕ' is semantically equivalent to ϕ (the negations of two semantically equivalent formulas are also semantically equivalent). Also, any logical operator in ϕ' is either the topmost $\neg$ or a logical operator in ψ' and so must equal either $\neg$ or $\wedge$ by the Induction Hypothesis. Therefore, all logical operators in ϕ' are from the set $\{\neg, \wedge\}$ as claimed.

Case 3: Let ϕ be obtained from operator $\wedge$ and so ϕ is equal to $\psi_1 \wedge \psi_2$ for some formulas ψ_1 and ψ_2. By the Induction Hypothesis, there exist formulas ψ_1' and ψ_2' such that $\psi_1' \equiv \psi_1$, $\psi_2' \equiv \psi_2$ and where all logical operators in ψ_1' and ψ_2' are from $\{\neg, \wedge\}$. Therefore, ϕ' defined as $\psi_1' \wedge \psi_2'$ is semantically equivalent to ϕ and also contains only operators from the set $\{\neg, \wedge\}$.

Case 4:   Let ϕ be obtained from operator $\vee$. Then ϕ equals $\psi_1 \vee \psi_2$ for some
formulas ψ_1 and ψ_2. By the Induction Hypothesis, there are formulas ψ_1'
and ψ_2' such that $\psi_1' \equiv \psi_1$, $\psi_2' \equiv \psi_2$, and all logical operators in ψ_1' and ψ_2'
are in the set $\{\neg, \wedge\}$. We define ϕ' to be $\neg(\neg\psi_1' \wedge \neg\psi_2')$. By one of the De
Morgan's Laws and since $\psi_i' \equiv \psi_i$ for all i in $\{1,2\}$ we infer that $\phi \equiv \phi'$.
It is also clear that all logical operators in ϕ' are from the set $\{\neg, \wedge\}$ by
construction of ϕ', as this is the case for ψ_1' and ψ_2'.

Case 5:   Let ϕ be obtained from operator $\Rightarrow$. Then ϕ equals $\psi_1 \Rightarrow \psi_2$ for
some formulas ψ_1 and ψ_2. By Induction Hypothesis, there are formulas ψ_1'
and ψ_2' such that $\psi_1' \equiv \psi_1$, $\psi_2' \equiv \psi_2$, and all logical operators in ψ_1' and in
ψ_2' are from the set $\{\neg, \wedge\}$. We then use the general semantic equivalence
$\mathscr{A} \Rightarrow \mathscr{B} \equiv \neg\mathscr{A} \vee \mathscr{B}$ and one of the De Morgan's Laws to infer that ϕ'
defined as $\neg(\psi_1' \wedge \neg\psi_2')$ is semantically equivalent to ϕ and that all of its
logical operators are in the set $\{\neg, \wedge\}$.

Case 6:   Finally, let ϕ be obtained from operator $\Leftrightarrow$. Then ϕ equals $\psi_1 \Leftrightarrow \psi_2$
for some formulas ψ_1 and ψ_2. By the Induction Hypothesis, there exist
formulas ψ_1' and ψ_2' such that $\psi_1' \equiv \psi_1$, $\psi_2' \equiv \psi_2$, and where all logical
operators in ψ_1' and in ψ_2' are from the set $\{\neg, \wedge\}$. Similarly to the previous
case, we appeal to a general semantic equivalence $\mathscr{A} \Leftrightarrow \mathscr{B} \equiv (\mathscr{A} \wedge \mathscr{B}) \vee$
$(\neg\mathscr{A} \wedge \neg\mathscr{B})$ and one of the De Morgan's Laws to infer that ϕ' defined
as $\neg(\neg(\psi_1' \wedge \psi_2') \wedge \neg(\neg\psi_1' \wedge \neg\psi_2'))$, is semantically equivalent to ϕ and
contains logical operators only from the set $\{\neg, \wedge\}$.

$\square$

The above proof by Structural Induction can be reinterpreted as defining an
algorithm that computes, for each formula ϕ, a concrete formula ϕ' satisfying
the claim of the theorem. We leave it as an exercise to state the corresponding
recursive definition for such an algorithm.

Incidentally, Structural Induction is the foundation of most automated theorem
provers [10], which are special programming environments in which formal proofs
can be executed and verified by a machine. We will dedicate most of Section 5.7 to
emphasizing the significance of this proof technique in Informatics:

• In Section 5.7.2 we will formally prove the Substitution Lemma using Structural
  Induction. This lemma may rightly be seen as a key result for the entire area of
  program verification and formal methods (see also Section 5.7.3).

• In Section 5.7.3, we sketch the seeds of formal program verification. The approaches developed in this chapter are essential for the reliable implementation
  and validation of today's safety-critical systems or of those parts of systems that
  require high assurance – be it for resiliency, privacy, or other system aspects.

- In Section 5.7.4, we demonstrate that the idea of Structural Induction may be generalized so that it is applicable to more complex and typically hierarchical structures found in Informatics.

5.6 Mathematical Induction

Historically, doing inductive proof is intimately connected with a special instance of Well-Founded Induction, the so-called *Mathematical Induction*. As we saw in Section 4.1, the Proof Principle of Mathematical Induction is directly derivable from the Induction Axiom P5 of the Axioms of Peano Arithmetic: for a proposition $\mathscr{A}$ of interest, we merely have to define M as the set of natural numbers that satisfy proposition $\mathscr{A}$. We may then prove that all natural numbers have property $\mathscr{A}$ by proving the proof obligation of Peano Axiom P5:

$$\forall M \subseteq \mathbb{N}.\, 0 \in M \,\wedge\, \forall n \in \mathbb{N}.\, n \in M \Rightarrow \mathfrak{s}(n) \in M \tag{5.14}$$

We may state this approach formally as a proof principle:

Proof Principle 13 (Principle of Mathematical Induction)

Suppose that we have a proposition $\mathscr{A}$ about natural numbers that is true for the natural number 0, and where the truth of $\mathscr{A}$ for a natural number $\mathfrak{s}(n)$ is derivable from the truth of $\mathscr{A}$ for its predecessor n. Then proposition $\mathscr{A}$ is true for all natural numbers. Formally:

$$\big(\mathscr{A}(0) \,\wedge\, \forall n \in \mathbb{N}.\, \mathscr{A}(n) \Rightarrow \mathscr{A}(n+1)\big) \;\Rightarrow\; \forall n \in \mathbb{N}.\, \mathscr{A}(n)$$

Note that Mathematical Induction is a special case of Structural Induction, where the inductively defined set is that of the natural numbers $\mathbb{N}$, the set of atoms is $\{0\}$, and the set of operators is $\{\mathfrak{s}(\cdot)\}$. We now apply Mathematical Induction to show some central properties of the addition and multiplication of natural numbers, based on the inductive definitions of these operations given in Definitions 4.2 and 4.3.

Theorem 5.5. *Let n, m, and k be in $\mathbb{N}$. Then we have:*

Associativity:
$$(n+m)+k \;=\; n+(m+k) \qquad\qquad (n\cdot m)\cdot k \;=\; n\cdot(m\cdot k)$$

Commutativity:
$$n+m \;=\; m+n \qquad\qquad\qquad n\cdot m \;=\; m\cdot n$$

Neutral Elements:
$$n+0 \;=\; n \qquad\qquad\qquad\quad n\cdot 1 \;=\; n$$

Right Cancellation Rules:
$$(n+k = m+k) \Rightarrow (n=m) \qquad (n \cdot k = m \cdot k) \wedge (k \neq 0) \Rightarrow (n=m)$$

Distributivity:
$$(n+m) \cdot k = n \cdot k + m \cdot k$$

Proof. We prove here the properties *Associativity* and *Commutativity* for the sake of illustration. The proofs for the remaining properties are left as exercises.

The proof of *Associativity* uses Mathematical Induction over the argument n, and so leaves m and k as arbitrary natural numbers. The proof of *Commutativity*, in contrast, needs to nest two instances of Mathematical Induction, an outermost one for n and an innermost one for m. It is worth nothing that such a nesting would not be required if we were to use Well-Founded Induction instead (see page 173 for details on that point).

Let us first prove the property *Associativity*. For that, we let m and k be arbitrary but fixed natural numbers. Note that Theorem 5.5 contains an implicit universal quantification over n, m, and k, and so here we eliminate that quantification for m and k, letting them be arbitrary but fixed natural numbers. We now use Mathematical Induction on the remaining parameter n. The property in question is $\mathscr{A}(n)$, saying that $(n+m)+k$ equals $n+(m+k)$. Note that $\mathscr{A}$ has only one parameter as k and m are arbitrary but fixed.

Induction Base Case: Let n be 0. Then we have:

$$(0+m)+k \stackrel{\text{(Def. 4.2.a)}}{=} m+k \stackrel{\text{(Def. 4.2.a)}}{=} 0+(m+k)$$

Induction Step: Let the property $\mathscr{A}(n)$, i.e., $(n+m)+k = n+(m+k)$, be true for an arbitrary but fixed natural number n; this is the Induction Hypothesis. We then need to show that the property $\mathscr{A}(n+1)$ is true:

$$((n+1)+m)+k \stackrel{\text{(Lemma 4.2)}}{=} (\mathfrak{s}(n)+m)+k \stackrel{\text{(Def. 4.2.b)}}{=} \mathfrak{s}(n+m)+k$$
$$\stackrel{\text{(Def. 4.2.b)}}{=} \mathfrak{s}((n+m)+k) \stackrel{\text{(IH)}}{=} \mathfrak{s}(n+(m+k))$$
$$\stackrel{\text{(Def. 4.2.b)}}{=} \mathfrak{s}(n)+(m+k) \stackrel{\text{(Lemma 4.2)}}{=} (n+1)+(m+k)$$

To show the property *Commutativity* for addition, we define the property $\mathscr{A}$ as having one parameter so that $\mathscr{A}(n)$ is true if and only if $n+m$ equals $m+n$ for *all* m in $\mathbb{N}$. We use Mathematical Induction over n to show that $\mathscr{A}(n)$ holds for all n in $\mathbb{N}$, and this then shows that addition is commutative:

Induction Base Case: This proceeds as in the case of $n=0$ for the alternative proof given on page 173.

Induction Step: Let n be an arbitrary but fixed element of $\mathbb{N}$ such that $\mathscr{A}(n)$ holds. We now need to show that $\mathscr{A}(n+1)$, i.e., $\mathscr{A}(\mathfrak{s}(n))$, holds as well – which we do as follows (where m is an arbitrary but fixed natural number):

$$(n+1)+m \overset{\text{(Lemma 4.2)}}{=} \mathfrak{s}(n)+m \overset{\text{(Def. 4.2.b)}}{=} \mathfrak{s}(n+m) \overset{\text{(IH)}}{=} \mathfrak{s}(m+n)$$
$$\overset{(*)}{=} m+\mathfrak{s}(n) \overset{\text{(Lemma 4.2)}}{=} m+(n+1)$$

The above reasoning appealed to an equation marked with $(*)$. We therefore owe a proof that this equation is valid in order for the above reasoning to be correct and complete. That is to say, we need to prove that

$$m+\mathfrak{s}(n) \;=\; \mathfrak{s}(m+n) \tag{$*$}$$

holds for all m and n in $\mathbb{N}$. We do this using, again, Mathematical Induction, but now over the parameter m. So let $\mathscr{A}(m)$ denote that $m+\mathfrak{s}(n)$ equals $\mathfrak{s}(m+n)$ for all n in $\mathbb{N}$. We have two cases to consider:

Induction Base Case: Let m be 0. We mean to show that $\mathscr{A}(0)$ is true. So let n be an arbitrary element of N. Then we have:

$$0+\mathfrak{s}(n) \overset{\text{(Def. 4.2.a)}}{=} \mathfrak{s}(n) \overset{\text{(Def. 4.2.a)}}{=} \mathfrak{s}(0+n)$$

Since n was an arbitrary natural number, this shows that $\mathscr{A}(0)$ is true.

Induction Step: Let $\mathscr{A}(m)$ be true for an arbitrary but fixed m in $\mathbb{N}$. Then we have to show that $\mathscr{A}(m+1)$ holds as well. To that end, let n be an arbitrary but fixed natural number. Then we have:

$$(m+1)+\mathfrak{s}(n) \overset{\text{(Lemma 4.2)}}{=} \mathfrak{s}(m)+\mathfrak{s}(n) \overset{\text{(Def. 4.2.b)}}{=} \mathfrak{s}(m+\mathfrak{s}(n)) \overset{\text{(IH)}}{=} \mathfrak{s}(\mathfrak{s}(m+n))$$
$$\overset{\text{(Def. 4.2.b)}}{=} \mathfrak{s}(\mathfrak{s}(m)+n) \overset{\text{(Lemma 4.2)}}{=} \mathfrak{s}((m+1)+n)$$

Since n was arbitrary, we infer from this that $\mathscr{A}(m+1)$ is true as well, as desired.
$\square$

Remark 5.1. In the above inductive reasoning, we marked all places in which the respective Induction Hypothesis was used with (IH). An inductive proof always has some places in which the Induction Hypothesis is used, and it is important that we clearly indicate such places – for example as done above.

We may apply the Principle of Mathematical Induction, in combination with the inductively defined operations over natural numbers, to prove propositions more complex than those in Theorem 5.5.

Example 5.5 (Examples of Mathematical Induction).
For all natural numbers n in $\mathbb{N}$ we have:

1. There are 2^n subsets of a set that has exactly n elements.

2. $\sum_{i=1}^{n} i = \frac{n(n+1)}{2}$ is a formula for the sum of the first n natural numbers.

3. $\sum_{i=1}^{n} (2i-1) = n^2$ is a formula for the sum of the first n odd natural numbers.

Proof. We only prove the first of these claims above, and leave the proofs of the remaining claims as exercises. The property $\mathscr{A}$ we want to prove, then, has parameter n where $\mathscr{A}(n)$ is true if and only if *all* sets of exactly n elements have 2^n subsets:

Induction Base Case: Let n be 0. A set that has exactly 0 elements has to be equal to the empty set, and the latter has exactly $1 = 2^0$ subsets, namely the empty set itself. Therefore, $\mathscr{A}(0)$ is true.

Induction Step: Let $\mathscr{A}(n)$ be true for an arbitrary but fixed natural number n. The truth of $\mathscr{A}(n)$ is of course the Induction Hypothesis. We need to show that $\mathscr{A}(n+1)$ is true. So let M be a set with exactly $n+1$ elements. It suffices to show that M has 2^{n+1} subsets. Since $n+1 > 0$, we know that M cannot be the empty set. Therefore, M contains at least one element, say m in M. We now use this element m to reduce the counting of subsets of M to the counting of subsets of sets of size n, for which the Induction Hypothesis holds! The key to obtaining this reduction is to classify subsets of M into those that contain m and those that don't. So let U be an arbitrary subset of M:

• Let m not be in U. Then U is a subset of the set $M \setminus \{m\}$, i.e., $U \subseteq M \setminus \{m\}$, which contains exactly $(n+1) - 1 = n$ elements. Conversely, any subset of $M \setminus \{m\}$ is a subset of M that does not contain m. By Induction Hypothesis, this shows that there are 2^n subsets of M that do not contain m.

• Let m be in U. Then, the set $U' =_{df} U \setminus \{m\}$, obtained from U be removing element m, is a subset of $M \setminus \{m\}$ by construction. Conversely, every subset U' of $M \setminus \{m\}$ corresponds to a unique subset U of M that contains m, based on $U \setminus \{m\} = U'$ or, equivalently, $U' \cup \{m\} = U$. Therefore, we infer from the Induction Hypothesis that there are 2^n subsets of M that contain m.

Clearly, a subset of M either contains element m or it does not. Therefore, the above analysis reveals that there are $2^n + 2^n = 2 \cdot 2^n = 2^{n+1}$ subsets of set M. Since M was an arbitrary set of $n+1$ elements, this proves that $\mathscr{A}(n+1)$ is true – as desired.   □

It is helpful to always keep in mind that Mathematical Induction is a special instance of Structural Induction, where the inductively defined set is $\mathbb{N}$, the sole atom is 0, and the sole operator is $\mathfrak{s}(\cdot)$.

5.7 Reflections: Exploring Covered Topics More Deeply

Mathematicians often readily employ the most powerful proof method available. For inductive proofs, for example, they may choose Well-Founded Induction or the equivalent Proof Principle of Minimal Counter Example (see Chapter 6), which combines well-founded reasoning with the Proof Principle of Proof by Contradiction. The more we are familiar with an induction principle we wish to apply, the better we can guide the required inductive proof that uses this principle. Such understanding sometimes even leads to the complete automation of such an inductive proof.

A central aspect of any inductive proof is the identification, and effective use, of a suitable Induction Hypothesis. As we will see in Section 5.7.1, it is not always sufficient to consider the claimed proposition $\mathscr{A}$ as Induction Hypothesis. In such cases, we need to *strengthen* the proposition we mean to show so that the strengthened version has a successful inductive proof, and logically implies the original proposition. We may think of such a strengthened proposition as another example of an Archimedean Point, a concept that often comes up in Informatics – especially in research on how to automatically discover suitable invariants in programs.

The remainder of this chapter will therefore first illustrate the need for strengthening the Induction Claim in general (Section 5.7.1) and then make an excursion into the area of program verification (Sections 5.7.2 and 5.7.3). Finally, we will discuss additional types of inductive proofs – motivated by practical considerations (Sections 5.7.4 and 5.7.5).

5.7.1 Strengthening of Induction Claims

Sometimes we cannot successfully apply an Induction Principle directly. When we analyze why the proof attempt failed, we then typically see that the Induction Hypothesis was too weak to logically imply the claim that would then complete the Induction Step. Exemplary for this situation is the existence of negation normal forms for formulas of propositional logic, as discussed on page 28. Let us recall that a formula is in *negation normal form* (NNF) if all of its negation symbols are applied directly to atomic propositions only. For example, formula $\neg \mathscr{A} \vee \mathscr{B}$ is in NNF whereas $\neg (\mathscr{A} \vee \mathscr{B})$ is not.

We were able to prove the functional completeness of the set of logical operators $\{\wedge, \neg\}$ for all formulas of propositional logic by using structural induction (see page 181). However, the task of proving that every formula is semantically equivalent to one in NNF is more difficult. If we were to mimic the proof on page 181, for example, we would consider a case ϕ equal to $\neg \psi$ where the Induction Hypothesis is that ψ is semantically equivalent to a formula ψ' that is in NNF. But this is insufficient, since then the formula $\neg\psi'$, as a candidate NNF semantically equivalent to ϕ, is semantically equivalent to ϕ but is *not* in NNF whenever ψ is not an atomic proposition. This means that Structural Induction is now stuck and we cannot successfully complete the proof! The key to this problem is a judicious strengthening of the Induction Claim. We illustrate this now through an example.

Example 5.6 (Existence of an NNF). We consider the set of *formulas of propositional logic*, as specified in Definition 2.2 (page 23). In our inductive proof, we appeal to the proved functional completeness of the set of logical operators $\{\neg, \wedge\}$ for all formulas of propositional logic (see Theorem 5.4 on page 181). Concretely, this means that it suffices to show that all formulas built using only $\neg$ and $\wedge$ have a semantically equivalent formula that is in NNF. Let us first see why this is the case: consider an *arbitrary* formula ϕ of propositional logic. By Theorem 5.4, there is a formula ψ that is semantically equivalent to ϕ and has as logical operators only those in $\{\neg, \wedge\}$. Suppose we can show that for such a ψ there is a formula η of propositional logic (regardless of whether its operators are from a special set or not) that is semantically equivalent to ψ and in NNF. Then η is also semantically equivalent to ϕ and (still) in NNF. To summarize, it suffices to prove the existence of semantically equivalent NNFs for formulas built out of atomic propositions, $\neg$, and $\wedge$ only.

As already suggested above, a direct attempt to inductively prove the Induction Claim

$\mathscr{E}_{NNF}$:   *Every formula of propositional logic whose logical operators are from the set $\{\neg, \wedge\}$ has a semantically equivalent formula of propositional logic in NNF*

through structural induction is too weak and will fail. But this problem goes away if we *strengthen* the Induction Claim to:

$\mathscr{E}_{NNF}^+$:   *Every formula ϕ of propositional logic whose logical operators are from the set $\{\neg, \wedge\}$ has formulas ϕ^+ and ϕ^- of propositional logic that are both in NNF such that $\phi \equiv \phi^+$ and $\neg\phi \equiv \phi^-$.*

The strengthening therefore consists of also constructing a semantically equivalent NNF for the negation of the formula in question. As we can see below, the proof of Induction Claim $\mathscr{E}_{NNF}^+$ now uses Structural Induction without any problems. Note that structural induction here only needs to consider three cases – atomic propositions, formulas that are negations, and formulas that are conjunctions – due to the fact that the claim is only for formulas from that fragment of propositional logic:

Case 1:   Let ϕ be an atomic proposition $\mathscr{A}$. Then we may set $\phi^+ =_{df} \phi$ and $\phi^- =_{df}$ $\neg\mathscr{A}$. These two formulas are clearly in NNF and semantically equivalent (in fact identical) to ϕ, respectively $\neg\phi$.

Case 2:   Let ϕ be a negation, so it equals $\neg\psi$ where ψ is a formula whose logical operators are from the set $\{\neg, \wedge\}$. By Induction Hypothesis, there are

formulas ψ^+ and ψ^- of propositional logic in NNF such that $\psi \equiv \psi^+$
and $\neg\psi \equiv \psi^-$. Then we may set $\phi^+ = \psi^-$ and $\phi^- = \psi^+$. Clearly, ϕ^+
and ϕ^- are in NNF with $\phi \equiv \phi^+$ and $\neg\phi = \neg\neg\psi \equiv \psi \equiv \psi^+ = \phi^-$. In
particular, we get $\neg\phi \equiv \phi^-$ as desired.

Case 3:   Let ϕ be a conjunction, so it equals $\phi_1 \wedge \phi_2$ for formulas ϕ_1 and ϕ_2 of
propositional logic whose logical operators are from the set $\{\neg, \wedge\}$. By
Induction Hypothesis, there are formulas ϕ_1^+, ϕ_1^-, ϕ_2^+, and ϕ_2^- in NNF
such that, for all i in $\{1,2\}$, we have $\phi_i \equiv \phi_i^+$ and $\neg\phi_i \equiv \phi_i^-$. We now
set $\phi^+ =_{df} \phi_1^+ \wedge \phi_2^+$ and $\phi^- =_{df} \phi_1^- \vee \phi_2^-$. Note that ϕ^- contains also the
logical operator $\vee$, which is permitted according to the Induction Claim
$\mathscr{E}_{NNF}^+$. It is clear from the Induction Hypothesis that ϕ^+ and ϕ^- are in
NNF since this property is closed under conjunctions and disjunctions.
We now reason that these definitions allow us to prove the two claims for
ϕ:

$$\phi = \phi_1 \wedge \phi_2 \equiv \phi_1^+ \wedge \phi_2^+ = \phi^+$$
$$\neg\phi = \neg(\phi_1 \wedge \phi_2) = \neg\phi_1 \vee \neg\phi_2 \equiv \phi_1^- \vee \phi_2^- = \phi^-$$

Therefore, we have proved this third case as well. $\square$

The determination of the right degree of strengthening, i.e., the identification of the
Archimedean Point that captures the problem at hand, is an artistic skill for which
there is no ready-made or foolproof recipe. But here the familiar motto applies as
well: *Skill comes with practice.*

5.7.2 Important Application: The Substitution Lemma

The *Substitution Lemma* provides a crucial relationship between the syntactic substitution introduced in Definition 4.6 (page 134) and its *semantic* counterpart. This
substitution lemma and many of its variants are the foundation of many formal proof
procedures. We will discuss this aspect in further detail in Section 5.7.3. We emphasize that the syntactic substitution as well as the semantics one defines below rest
on the Boolean terms introduced in Definition 4.5 (page 133), not on definitions of
formulas of propositional logic (as considered in the last example).

First, we need to define the semantic counterpart of syntactic substitution.

Definition 5.7 (Semantic substitution). Let β be an element of $\mathscr{B}_{\mathscr{V}}$, i.e., a function $\beta\colon \mathscr{V} \to \{tt, ff\}$ that associates variables from a set of variables $\mathscr{V}$ with truth
values. The *semantic substitution* $\cdot\{\cdot \mapsto \cdot\}\colon \mathscr{B}_{\mathscr{V}} \times \mathscr{V} \times \{tt, ff\} \to \mathscr{B}_{\mathscr{V}}$ is defined
as follows, for all v in $\{tt, ff\}$, and X and Y in $\mathscr{V}$:

$$\beta\{X \mapsto v\}(Y) =_{df} \begin{cases} v & \text{if } X = Y, \\ \beta(Y) & \text{otherwise} \end{cases}$$

$\square$

Put into words, the application of $\{X \mapsto v\}$ on β renders a new assignment that is the same as β except that at variable X it has truth value v. We may now put the semantic and syntactic substitutions into an important relationship.

Lemma 5.2 (Substitution Lemma). *Let t and t' be in $\mathscr{BT}$, Boolean terms over a variable set $\mathscr{V}$. Let X be a variable in $\mathscr{V}$ and β in $\mathscr{B}_{\mathscr{V}}$ an assignment. Then we have:*

$$[\![t[X \mapsto t']]\!]_B(\beta) = [\![t]\!]_B(\beta\{X \mapsto [\![t']\!]_B(\beta)\}) \tag{5.15}$$

Let us translate this mathematical fact into words: for an arbitrary but fixed assignment β, the syntactic substitution of a variable X with a term t' has the same effect as the semantic substitution of variable X with the (semantic) value of t' with respect to β. This correspondence is the typical effect we expect, for example, in the assignment statements of conventional programming languages.

The proof of this lemma is rather long, but most of its length is due to administrative overhead that comes from the rigorous and repeated application of Definition 4.10 (definition of $[\![\cdot]\!]_B$, page 148), Definition 4.6 (definition of syntactic substitution, page 134), and Definition 5.7 (definition of semantic substitution). The few places in which we invoke the Induction Hypothesis don't add much to that complexity.

Proof. We proceed by Structural Induction over the term t in (5.15).

Case 1:   Let t equal T. Then we have

$$[\![\mathsf{T}[X \mapsto t']]\!]_B(\beta) \overset{\text{(Def. 4.6)}}{=} [\![\mathsf{T}]\!]_B(\beta)$$
$$\overset{\text{(Def. 4.10)}}{=} tt$$
$$\overset{\text{(Def. 4.10)}}{=} [\![\mathsf{T}]\!]_B(\beta\{X \mapsto [\![t']\!]_B(\beta)\})$$

Case 2:   Let t equal F. Then we reason analogously to Case 1 (which we leave as an exercise).

Case 3:   Let t equal Y for some Y in $\mathscr{V}$ where $Y \neq X$.

$$[\![Y[X \mapsto t']]\!]_B(\beta) \overset{\text{(Def. 4.6)}}{=} [\![Y]\!]_B(\beta)$$
$$\overset{\text{(Def. 4.10)}}{=} \beta(Y)$$
$$\overset{\text{(Def. 5.7)}}{=} \beta\{X \mapsto [\![t']\!]_B(\beta)\}(Y)$$

Case 4:   Let t equal X.

$$[\![X[X \mapsto t']]\!]_B(\beta) \overset{\text{(Def. 4.6)}}{=} [\![t']\!]_B(\beta)$$

$$\overset{\text{(Def. 5.7)}}{=} \beta\{X \mapsto [\![t']\!]_B(\beta)\}(X)$$

$$\overset{\text{(Def. 4.10)}}{=} [\![X]\!]_B\Big(\beta\{X \mapsto [\![t']\!]_B(\beta)\}\Big)$$

Case 5: Let t equal $\neg t_1$ for some Boolean term t_1.

$$[\![(\neg t_1)[X \mapsto t']]\!]_B(\beta) \overset{\text{(Def. 4.6)}}{=} [\![\neg t_1[X \mapsto t']]\!]_B(\beta)$$

$$\overset{\text{(Def. 4.10)}}{=} \dot{\neg}\Big([\![t_1[X \mapsto t']]\!]_B(\beta)\Big)$$

$$\overset{\text{(IH)}}{=} \dot{\neg}\Big([\![t_1]\!]_B(\beta\{X \mapsto [\![t']\!]_B(\beta)\})\Big)$$

$$\overset{\text{(Def. 4.10)}}{=} [\![\neg t_1]\!]_B(\beta\{X \mapsto [\![t']\!]_B(\beta)\})$$

Case 6: Let t equal $(t_1 \vee t_2)$ for some Boolean terms t_1 and t_2.

$$[\![(t_1 \vee t_2)[X \mapsto t']]\!]_B(\beta) \overset{\text{(Def. 4.6)}}{=} [\![(t_1[X \mapsto t'] \vee t_2[X \mapsto t'])]\!]_B(\beta)$$

$$\overset{\text{(Def. 4.10)}}{=} ([\![t_1[X \mapsto t']]\!]_B(\beta)\dot{\vee}[\![t_2[X \mapsto t']]\!]_B(\beta))$$

$$\overset{\text{(IH)}}{=} ([\![t_1]\!]_B(\beta\{X \mapsto [\![t']\!]_B(\beta)\})\dot{\vee}[\![t_2]\!]_B(\beta\{X \mapsto [\![t']\!]_B(\beta)\}))$$

$$\overset{\text{(Def. 4.10)}}{=} [\![(t_1 \vee t_2)]\!]_B(\beta\{X \mapsto [\![t']\!]_B(\beta)\})$$

Case 7: Let t equal $(t_1 \wedge t_2)$ for some Boolean terms t_1 and t_2. Then we reason analogously to Case 6 (which we leave as an exercise).   □

It is worth the time to look at each of the above cases in detail, and to check and understand why all equations in those cases are valid. This inspection will reveal that most of the above reasoning is the *literal* application of the corresponding definitions. Those reasoning steps that are annotated with (IH) appeal to the Induction Hypothesis, that is to instances of the Substitution Lemma that involve genuine subterms of the currently handled Boolean term, in order to validate those equations.

5.7.3 Formal Correctness Proofs

The Substitution Lemma, discussed in Section 5.7.2 above, and its variants are of fundamental importance for many formal proof procedures. This becomes particularly apparent when we consider the rules for axiomatic proofs in propositional logic. As already mentioned in Section 2.3.2, all valid semantic equivalences can be derived by the application of the rules that are above the horizontal line in Lemma 2.1; no other rules are needed for this. Central to this result is the Principle of *Substitution of Equals for Equals*. If we study this principle now in the formal model of propositional logic given by Boolean terms, then the method of axiomatic proof ultimately derives its justification from the formal result stated in Theorem 5.6 below. We should recall here that we defined the semantic equivalence $\equiv$ on Boolean terms through an inductively defined semantics $[\![\cdot]\!]_B$ (see Definition 4.11 on page 149).

Theorem 5.6 (Compositionality of $[\![\cdot]\!]_B$). *Let t, t', and t'' be in $\mathscr{B}\mathscr{T}$ such that $t' \equiv t''$. Further, let X be a variable in $\mathscr{V}$. Then we have*

$$t[X \mapsto t'] \equiv t[X \mapsto t''] \tag{5.16}$$

That is to say, we may simultaneously replace equals with (semantic) equals.

Proof. The proof is quite simple, as we can invoke the Substitution Lemma. Let t, t', t'' be in $\mathscr{B}\mathscr{T}$, let $t' \equiv t''$, and let X be in $\mathscr{V}$. Then we have:

$$
\begin{aligned}
[\![t[X \mapsto t']]\!]_B(\beta) &= [\![t]\!]_B(\beta\{X \mapsto [\![t']\!]_B(\beta)\}) && \text{(Substitution Lemma 5.2)} \\
&= [\![t]\!]_B(\beta\{X \mapsto [\![t'']\!]_B(\beta)\}) && (t' \equiv t'') \\
&= [\![t[X \mapsto t'']]\!]_B(\beta) && \text{(Substitution Lemma 5.2)}
\end{aligned}
$$
$\square$

The above application of the Substitution Lemma in propositional logic made use of variables as a kind of placeholder for arbitrary Boolean terms. In Informatics, variables also play a central role as a key ingredient of programming languages – notably those with *imperative* features that explicitly maintain and manipulate computational state. In imperative languages, variables act as receptacles or *containers*. An executing program may then place concrete values into such containers or replace values already stored there with new ones. In fact, we may interpret the meaning of an imperative program as its input/output behavior: how the initial values in such containers change once the program has stopped its execution.

Program Verification [5, 25] is an area of Informatics that develops methods for formally proving properties of programs, for example that an imperative program has a specified input/output behavior (see Section 5.3.1). A pioneering contribution in that regard was the so-called *Hoare Calculus*. It allows us to prove the correctness of programs with a rule-based, formal proof that is compositional in the structure of the program. The correctness properties that we wish to prove in the Hoare Calculus are expressed in the form of Hoare triples, which we introduced on page 177. Formal proofs, including correctness proofs within the Hoare Calculus, essentially perform a purely syntactic manipulation of formulas – in contrast to semantic reasoning of the kind we discussed for equivalences of propositional logic. These syntactic manipulations reflect the composition of a program but also produce *proof obligations*, which are formulas in a mathematical logic that we need to show to be tautologies. Therefore, a Hoare Calculus not only consists of its own proof rules for program constructs but also depends on the ability to prove tautologies in a mathematical logic. As a consequence, automated program verification also depends on automated theorem provers. Much research progress has been made in automated program verification, and large companies such as Intel, Microsoft, Google, and Facebook are now actively developing program verification tools to satisfy in-house needs for software verification – see for example the research reported in [14, 13, 21, 9].

Remember, in the Hoare Calculus, a Hoare triple $\{P\}$ *prog* $\{Q\}$ denotes a program *prog* and formulas P and Q of mathematical logic. The intuition is that formulas P and Q describe a set of computational states of the program. For example P denotes the set of states that *satisfy* formula P. Formula $(x > 5) \wedge (x + y = 1)$, for example, describes the set of all states where the value stored in the container for x is larger than 5 and the sum of the values stored in the containers for x and y equals 1. The intuitive meaning of a Hoare triple $\{P\}$ *prog* $\{Q\}$ is then that program *prog* outputs only states satisfying Q whenever it has an input state that satisfies P. Formula P is called the *precondition* and formula Q the *postcondition* of that Hoare triple.

A central operation in imperative programming is the assignment statement, in which a variable is assigned the meaning of an expression. Hoare Calculus has a proof rule that captures the semantics of such assignments syntactically:

$$ASS \ \frac{-}{\{Q[x \mapsto t]\} \, x = t \, \{Q\}} \tag{5.17}$$

It is best to read the conclusion of this proof rule (below the bar) from right to left. The execution of the assignment $x = t$ results in a state satisfying Q if this execution is done in a state satisfying $Q[x \mapsto t]$. In other words, if the execution starts in a state that satisfies Q when all occurrences of x in Q are interpreted as expression t (which is what formula $Q[x \mapsto t]$ represents), then the assignment $x = t$ puts the value of expression t into the container for x (and does nothing else) and so this results in a state satisfying Q. This rule may seem counter-intuitive at first sight; we leave it as an exercise to reason informally why the Hoare triple

$$\{x * y > y\} \, x = x * y \, \{x > y\} \tag{5.18}$$

is valid in this sense.

The Hoare Calculus has a much more expressive language for logical formulas than that for Boolean terms. Such formulas may contain program variables that are put into relation with other variables or constants using equality $=$, inequality $>$, and so forth. One such example is program variable x in (5.18) above. Strictly speaking, we need to extend the inductive definition of $[x \mapsto t]$ from Boolean terms to formulas of that logic, which may also contain universal and existential quantifiers.

The proof rules for Hoare logic have zero or more *premises* written on the top of the bar and one *conclusion* written below the bar. The intuition is that a rule can be invoked if all its premises are true (i.e., have been proved already), in which case its conclusion is also true. The proof rule *ASS* for Assignment in (5.17) is a special case of this format; the "−" above the rule bar indicates that there are *no* premises and so the conclusion is always true. Rules without premises are called *axioms*.

Since imperative programs can ultimately only change state through the execution of Assignment statements, we may view proof rule *ASS* rightly as taking a central place in all Hoare Calculi that are used to verify the correctness of imperative programs.

The Substitution Lemma, when extended to the formulas of the mathematical logic used in a given Hoare Calculus, provides the core of the argument that the proof rule *ASS* itself is correct, meaning that all conclusions of *ASS* are valid statements about the input/output behavior of assignments. Technically, let β be a function that represents a state: for each variable x of the program in question $\beta(x)$ is the value of variable x at "state" β. The Substitution Lemma then says that whenever formula $Q[x \mapsto t]$ is true at state β, then Q is true in the state that is the same as β except that variable x now has the value $[\![t]\!](\beta)$ that expression t has in state β.

Let us now try to prove the correctness of the Hoare triple

$$\{x = 10\}\, x = x + 3 \,\{x > 10\}$$

It appears that this is not directly possible through an application of proof rule *ASS*, since the two formulas do not match the pattern of that proof rule. Rather, we can use rule *ASS* on the formula $x > 10$ to infer the formula that we need at the front of the Hoare triple:

$$ASS \; \frac{-}{\{x + 3 > 10\}\, x = x + 3 \,\{x > 10\}}$$

Fortunately, this is a crucial step in proving the correctness of the Hoare triple $\{x = 10\}\, x = x + 3 \,\{x > 10\}$. We only have to reason that $x + 3 > 10$ is a logical consequence of $x = 10$, i.e., that the precondition of $\{x = 10\}\, x = x + 3 \,\{x > 10\}$ implies the precondition of $\{x + 3 > 10\}\, x = x + 3 \,\{x > 10\}$.

The formula $(x = 10) \Rightarrow (x + 3 > 10)$ is thus an example of the aforementioned proof obligations, tautologies whose validity we need to prove in a proof system for the logic in which these formulas are specified. For formula $(x = 10) \Rightarrow (x + 3 > 10)$ this is a proof system for arithmetic with equality and the strict ordering. The Hoare Calculus interfaces with proofs of validity for mathematical formulas through its proof rule *CONS* for *logical consequence*:

$$CONS \; \frac{P \Rightarrow P' \quad \{P'\}\, S \,\{Q'\} \quad Q' \Rightarrow Q}{\{P\}\, S \,\{Q\}} \tag{5.19}$$

We may read this rule as the ability to strengthen the precondition of a Hoare triple and to weaken its postcondition. Proof rule *CONS* is not an axiom. To invoke it, we need to prove three things: the correctness of one Hoare triple (in the middle) and the validity of two logical consequences. This then allows us to infer the correctness of the Hoare triple below the bar of the rule. For example, we may use proof rule *CONS* to prove the correctness of the Hoare triple $\{x = 10\}\, x = x + 3 \,\{x > 10\}$ as follows:

$$CONS \; \frac{(x=10) \Rightarrow (x+3>10) \qquad ASS \; \overline{\{x+3>10\}\, x=x+3\, \{x>10\}} \qquad (x>10) \Rightarrow (x>10)}{\{x=10\}\, x=x+3\, \{x>10\}}$$

It should be intuitive that such proofs have the form of trees, where the root is at the bottom – the Hoare triple whose correctness we prove – and with leaves that either have no premise (for proof rule *ASS*) or state the validity of a logical consequence (for proof rule *CONS*). Strictly speaking, the proof above is only complete if we also provide mathematical proofs of the two logical consequences. For the second one, this is trivial as $\phi \Rightarrow \phi$ is always valid for all formulas ϕ. In fact, it is convenient to have two variants of rule *CONS* where either the precondition or the postcondition does not change.

Imperative programs increase their expressive power through the sequential composition of assignments or programs in general. This requires a proof rule that supports reasoning about such compositions, a so-called *Composition Rule*:

$$COMP \; \frac{\{P\}\, S_1\, \{R\} \quad \{R\}\, S_2\, \{Q\}}{\{P\}\, S_1;S_2\, \{Q\}} \tag{5.20}$$

It is striking that this proof rule resembles the meaning of relational composition that we discussed on page 71. To establish that P is a correct precondition of program $S_1;S_2$ for postcondition Q, we need an *intermediary* condition R that acts as precondition for S_2 and its postcondition Q, but that also acts as postcondition for the program S_1 and its precondition P. In practice, we can often compute a precondition from a program and its given postcondition. For rule *COMP* we would therefore compute R from Q and S_2 and then prove that the Hoare triple $\{P\}\, S_1\, \{R\}$ is correct.

Let us illustrate the use of this rule for the composition of assignments

$$x = x+y; \; y = x-y; \; x = x-y;$$

It should be noted that these assignments update variables x and y with expressions that refer to these variables. This makes it more difficult to express a precondition of this composed program, as we mean to capture the *initial* values of x and y in that computation. Fortunately, we may use equality and so-called *logical* variables x_0 and y_0 that don't denote program variables to capture and maintain such initial values in a precondition $(x = x_0) \land (y = y_0)$. Suppose we want to show that this composed program swaps the values of x and y, without resorting to any auxiliary program variable. Then we can write this as a Hoare triple

$$\{x = x_0 \land y = y_0\}\, x = x+y; \; y = x-y; \; x = x-y\, \{x = y_0 \land y = x_0\} \tag{5.21}$$

We may now use the Hoare Calculus to prove the correctness of this Hoare triple:

$$\text{CONS} \cfrac{pre \Rightarrow pre_3 \qquad \text{COMP} \cfrac{\text{ASS} \cfrac{\overline{\quad}}{\{pre_3\}\, x = x+y\, \{pre_2\}} \qquad \text{COMP} \cfrac{\text{ASS} \cfrac{\overline{\quad}}{\{pre_2\}\, y = x-y\, \{pre_1\}} \qquad \text{ASS} \cfrac{\overline{\quad}}{\{pre_1\}\, x = x-y\, \{post\}}}{\{pre_2\}\, y = x-y; x = x-y\, \{post\}}}{\{pre_3\}\, x = x+y; y = x-y; x = x-y\, \{post\}} \qquad post \Rightarrow post}{\{pre\}\, x = x+y; y = x-y; x = x-y\, \{post\}}$$

For the sake of compactness, this proof uses names for conditions whose meaning is defined here:

$$pre =_{df} x = x_0 \wedge y = y_0$$
$$post =_{df} x = y_0 \wedge y = x_0$$
$$pre_1 =_{df} x - y = y_0 \wedge y = x_0$$
$$pre_2 =_{df} x - (x - y) = y_0 \wedge (x - y) = x_0$$
$$pre_3 =_{df} (x + y) - ((x + y) - y) = y_0 \wedge ((x + y) - y) = x_0$$

It is easy to see that this proof requires only one proof of logical consequence that is non-trivial, meaning that it is not of the form $\phi \Rightarrow \phi$. It is the formula $pre \Rightarrow pre_3$, i.e., the logical consequence

$$(x = x_0 \wedge y = y_0) \Rightarrow ((x+y) - ((x+y) - y) = y_0 \wedge ((x+y) - y) = x_0)$$

The corresponding proof using the familiar axioms of arithmetic is left as an exercise.

The proof rules above are for a programming language that is quite limited: it can only express the sequential composition of assignment statements. Theoretical Informatics has developed methods that allow us to understand the expressive power of such programming languages. For example, such results imply that the language above cannot compute many important input/output behaviors of real variables as they occur in practice. In programming languages that are used in practice, therefore, we have further control structures such as if-statements, loops, procedures, threads, and so forth. Although it is not too hard to design proof rules for if-statements and loops, programming constructs for parallel computation – such as threads – make compositional reasoning and thus the design of Hoare proof rules challenging. Another dimension of complexity is the richness of data structures in mainstream programming languages, which support not only variables for real numbers and Booleans but also variables that represent arrays, or trees, or that point to other objects in the program heap space. We refer to the specialist literature for more details on such extensions of Hoare logic, see, e.g., [5, 25]. To get an impression of the use of automated correctness proofs of programs in practice, we recommend reading [10, 8].

5.7.4 BNF-Based Induction

In Section 4.3.3, we saw how BNFs support the elegant inductive definition of structures such as binary trees and Boolean terms. Upon closer inspection, we realize that the well-founded ordering defined on such structures expresses the notion of a subtree (an interpretation that also applies to Boolean terms as we can think of them syntactically as trees). But it turns out that BNFs and, more generally, Chomsky grammars [35] are not restricted to well-founded orderings captured by a sub-tree relationship. Rather, they rely on a quite universal method for the finite description of (infinite) languages. By a language we mean here a specific, typically infinite, set of sequences of letters; we referred to such sequences of letters as words already.

In fact, it is possible to apply the Proof Principle of *Course Of Values Induction* (see page 178) in a specific usage pattern that allows us to prove properties of all words in such a language (see also Section 5.7.5). This also illustrates how Course Of Values Induction can be used to manage several different Induction Hypotheses, where each one is an inductive invariant for a specific case, such that their combination leads to a global invariant that applies to all elements in the set of interest.

Let us demonstrate concretely how this can be done in the case of BNFs. For each non-terminal symbol we have an induction hypothesis stating what words can be generated with that non-terminal symbol. This fits well the reasoning about programs that assign not only arithmetic terms to variables but more structured objects such as trees. We may think now of a BNF as a *proof skeleton*, to which these induction hypotheses are *attached*. Such an annotated skeleton is then often the basis for a fully automated or at least partly automated induction proof.

Let us consider now a BNF with set of terminal symbols $\mathbf{T}$ and set of non-terminal symbols $\mathbf{N}$. Furthermore, for each non-terminal symbol $\langle X \rangle$ from $\mathbf{N}$ we have a predicate $\mathscr{A}_{\langle X \rangle}$ that specifies a set of words over $\mathbf{T}$ as the *inductive invariant* for non-terminal $\langle X \rangle$. The intuition is that this invariant formulates a property that holds of all words over terminal symbols from $\mathbf{T}$ that the BNF can generate from $\langle X \rangle$. We may write such a predicate as a function of type $\mathscr{A}_{\langle X \rangle} : \mathbf{T}^* \to \{\, tt, ff \,\}$.

We also need to formalize the set of words over $\mathbf{T}$ that the BNF can derive from $\langle X \rangle$ in n steps, as this will allow us to use Course Of Values Induction:

$$L_{\langle X \rangle}^n \ =_{df} \ \{ w \in \mathbf{T}^* \mid \langle X \rangle \Rightarrow^n w \} \tag{5.22}$$

It is worth noting that $L_{\langle X \rangle}^n$ may well be empty: it may require more than n steps to generate from $\langle X \rangle$ a word that does not contain any non-terminal symbols. For example, $L_{\langle X \rangle}^0$ is always empty as $\langle X \rangle$ is a non-terminal symbol.

With this formalization of $L_{\langle X \rangle}^n$ and the inductive invariants $\mathscr{A}_{\langle X \rangle}$ for each $\langle X \rangle$ in $\mathbf{N}$ at hand, we can now express the global invariant that ties together these inductive invariants. Since $L_{\langle X \rangle}^n$ records the derivation length, the global invariant shares this parameter as follows:

$$\mathscr{A}(n) \ =_{df} \ \forall \langle X \rangle \in \mathbf{N}. \ \forall w \in L_{\langle X \rangle}^n. \ \mathscr{A}_{\langle X \rangle}(w) \tag{5.23}$$

It is clear that $\mathscr{A}(n)$ is a predicate over the set of natural numbers $\mathbb{N}$. Therefore, we may prove that $\mathscr{A}(n)$ holds for all n in $\mathbb{N}$ using Course of Values Induction. First we formulate a proof principle and its corresponding proof obligation for such proofs.

Proof Principle 14 (Principle of Generalized Structural Induction)

Given a BNF as above, and a predicate $\mathscr{A}$ as defined in (5.23). Assume that, for each <X> in $\mathbf{N}$, the predicate $\mathscr{A}_{<X>}$ is indeed an inductive invariant: if all words generated from <X> in a derivation of length less than n have property $\mathscr{A}_{<X>}$, then this is also true for all words generated from <X> in a derivation of length n. Then, for all <X> in $\mathbf{N}$, all words generated by <X> have property $\mathscr{A}_{<X>}$.

We may write this succinctly using the global invariant $\mathscr{A}$ as:

$$\left(\forall n \in \mathbb{N}.\ (\forall m \in \mathbb{N}.\ m < n \Rightarrow \mathscr{A}(m)) \Rightarrow \mathscr{A}(n)\right) \Rightarrow \forall n \in \mathbb{N}.\ \mathscr{A}(n)$$

In practice, this Principle of Generalized Structural Induction is realized by proving the following proof obligation:

Proof Obligation 1 (For Generalized Structural Induction)

Let us make the assumptions made in Proof Principle 14. Under these assumptions, we prove for each production rule of the BNF, i.e., for each

$$<X> ::= w_0 <Y_1> w_1 \ldots <Y_k> w_k \tag{5.24}$$

with $k \geq 0$, $w_i \in \mathbf{T}^*$ for all i in $\{0,\ldots,k\}$, and $<Y_j> \in \mathbf{N}$ for all j in $\{1,\ldots,k\}$ the following:

For all j in $\{1,\ldots,k\}$ and $\tilde{w}_i$ in $\mathbf{T}^*$ such that $\mathscr{A}_{<Y_j>}(\tilde{w}_i)$ holds, we have that $\mathscr{A}_{<X>}(w_0\,\tilde{w}_1\,w_1\ldots\tilde{w}_k\,w_k)$ holds as well.

We note that the production rule in (5.24) does not have any non-terminal symbols on its right-hand side when k equals 0. In that case, we need to show $\mathscr{A}_{<X>}(w_0)$ as an Induction Base Case, without relying on any Induction Hypotheses. We can now show that proving this proof obligation is sufficient for proving generalized structural induction.

Theorem 5.7. *Suppose that we have proved Proof Obligation 1. Then the property $\mathscr{A}_{<X>}(w)$ is true for each non-terminal symbol <X> in $\mathbf{N}$ and for every word w in $\mathbf{T}^*$ that can be derived from <X> in its defining BNF.*

Proof. By definition of the predicate $\mathscr{A}(\cdot)$ in (5.23), it suffices to show that $\forall n \in \mathbb{N}.\ \mathscr{A}(n)$ holds. Based on the Proof Principle of Generalized Structural Induction (see Proof Principle 14), it suffices to show that the Induction Step

$$\forall n \in \mathbb{N}. \ (\forall m \in \mathbb{N}. \ m < n \ \Rightarrow \ \mathscr{A}(m)) \ \Rightarrow \ \mathscr{A}(n)$$

is valid. To that end, let n be in $\mathbb{N}$ such that $\mathscr{A}(m)$ holds for all $m < n$. We now have to show that $\mathscr{A}(n)$ holds as well. Let $<X>$ be in $\mathbf{N}$ and w in $L^n_{<X>}$; so w is a word of terminal symbols that is derivable from $<X>$ in n steps. Without loss of generality we may assume that $n \geq 1$, since $L^0_{<X>}$ is empty. We now proceed by mathematical induction:

- **Base Case:** Let n equal 1. Then there is a one-step derivation $<X> \Rightarrow w$. In particular, this means that the BNF must have the production rule $<X> ::= w$. In Proof Obligation 1, this case therefore corresponds to considering a production in (5.24) with $k = 0$, for which therefore $\mathscr{A}_{<X>}(w)$ has already been shown directly in the proof of that Proof Obligation.
- **Induction Step:** Let $n > 1$. Then the n-step derivation of w from $<X>$ can be written as a one-step derivation $<X> \Rightarrow \alpha$ followed by a $(n-1)$-step derivation $\alpha \Rightarrow^{n-1} w$:

$$<X> \ \Rightarrow \ \alpha \ \Rightarrow^{n-1} \ w \tag{5.25}$$

Clearly, α must contain at least one non-terminal symbol, say, $<Y>$. The $(n-1)$-step derivation in (5.25) then implicitly induces an m-step derivation $<Y> \Rightarrow^m w'$ of at most $n-1$ steps such that w' is a subword of w. In particular, $m < n$ follows. Let us summarize the findings of this analysis: there are words α_1 and α_2 in $(\mathbf{N} \cup \mathbf{T})^*$, as well as words w_1 and w_2 in $\mathbf{T}^*$ such that

$$<X> \ \Rightarrow \ \underbrace{\alpha_1 <Y> \alpha_2}_{\alpha} \ \Rightarrow^* \ w_1 <Y> w_2 \ \Rightarrow^m \ \underbrace{w_1 \, w' \, w_2}_{w} \tag{5.26}$$

By our Induction Hypothesis, we have that $\mathscr{A}_{<Y>}(w')$ is true. But this argument was made for an arbitrary non-terminal $<Y>$ in α and so it is valid for all such non-terminals. Therefore, we may apply Proof Obligation 1 to the above production rule $<X> ::= \alpha$ to infer, as desired, that $\mathscr{A}_{<X>}(w)$ is true. $\qquad \square$

We illustrate the application of Theorem 5.7 through a very simple example, the BNF

$$<A> ::= a<B>$$
$$<B> ::= b \mid b<A> \mid <A>b<A>$$

We mean to show that the non-terminal symbol $<A>$ derives only words in which there is an equal number of terminals a and b. To that end, we define the following non-homogeneous predicates for the two non-terminal symbols of this BNF:

- $\mathscr{A}_{<A>}(w)$: w contains as many letters a as it contains letters b.

- $\mathscr{A}_{<B>}(w)$: w contains exactly one more letter b than it contains letters a.

Proof. Based on Proof Obligation 1, we have a case analysis over all production rules of this BNF:

Case 1:   Production rule <A> ::= a<B>.
          Let $\tilde{w}$ be such that $\mathscr{A}_{<B>}(\tilde{w})$ is true. Then $\tilde{w}$ contains exactly one more
          b than a. Therefore, the word $a\tilde{w}$ has exactly as many letters a as it has
          letters b. But then we know that $\mathscr{A}_{<A>}(a\tilde{w})$ is true.

Case 2:   Production rule <B> ::= b.
          We immediately have that $\mathscr{A}_{<B>}(b)$ is true.

Case 3:   Production rule <B> ::= b<A>.
          Let $\tilde{w}$ be such that $\mathscr{A}_{<A>}(\tilde{w})$ is true. Then we know that $\tilde{w}$ contains ex-
          actly as many letters a as it contains letters b. Therefore, the word $b\tilde{w}$
          has exactly one more letter b than it has letters a. But then we have that
          $\mathscr{A}_{<B>}(b\tilde{w})$ is true.

Case 4:   Production rule <B> ::= <A>b<A>.
          Let $\tilde{w}_1$ and $\tilde{w}_2$ be such that $\mathscr{A}_{<A>}(\tilde{w}_i)$ is true for $i = 1, 2$. Then the words
          $\tilde{w}_1$ and $\tilde{w}_2$ each contain exactly as many letters a as they contain letters
          b. Therefore, the word $\tilde{w}_1 b \tilde{w}_2$ contains exactly one more letter b than it
          contains letters a. But then we know that $\mathscr{A}_{<B>}(\tilde{w}_1 b \tilde{w}_2)$ is true.   □

5.7.5 *Additional Forms of Inductive Proof*

The BNFs introduced in Section 4.3.3 are a special form of the so-called *Chom-sky Grammars* [35]. BNFs are *context-free* grammars: production rules of the form in (5.24) have a left-hand side that consists of a non-terminal symbol <X> only. Chomsky Grammars are more general as they may provide context for such non-terminals. This can be seen in the following type hierarchy, where the type numbers increase with increased specialization:

- **Type-0 Grammars:** Production rules have right-hand sides as in (5.24). But the left-hand sides are now finite words over non-terminal and terminal letters such that at least one non-terminal symbol occurs.
- **Type-1 Grammars:** These are also known as *monotone grammars*. They restrict Type-0 grammars so that all production rules (except one that may generate the empty word) have left-hand sides that are not longer (as words) than their right-hand side.
- **Type-2 Grammars:** These are the BNFs, which have production rules of the form shown in (5.24), where left-hand sides consist of a sole non-terminal only.
- **Type-3 Grammars:** This is the most restrictive type of grammar, also known as *right regular grammars*. It restricts BNFs so that all right-hand sides of rules may contain *at most one* non-terminal symbol which has to occur at the rightmost position.

The basic concepts of BNFs, such as the derivation relation, are the same though for these other types of grammars. The above Chomsky Hierarchy is important in

Informatics, with applications in compilers, verification, and so forth. It also has interesting implications with regard to computability, expressive power, and efficiency in the areas of formal language theory and (partly) automata theory.

We will not require formal definitions of these grammar types but we exploit that the derivation of words can be defined uniformly for all these types; in other words, the definition of such a derivation relation for Type-0 Grammars specializes to the other grammar types. This fact allows us to extend the principle of BNF-based induction, introduced in Section 5.7.4, to obtain further grammar-based induction principles by a simple adjustment: Instead of defining individual Induction Hypotheses for each non-terminal symbol <X>, Induction Hypotheses have now to be defined for all left-hand sides of all production rules of the grammar. We illustrate this extended proof principle for a monotone grammar, by showing that it generates the language

$$\mathscr{L} =_{df} \{a^n b^n c^n \mid n \in \mathbb{N}\} \tag{5.27}$$

This language is the classical example that is used to prove that Type-1 Grammars are more expressive than Type-2 Grammars. We will show that the following monotone grammar generates this language.

$$
\begin{aligned}
\text{<S>} &::= \varepsilon \mid a\text{<S><B><C>} \\
\text{<C><B>} &::= \text{<B><C>} \\
a\text{<B>} &::= ab \\
b\text{<B>} &::= bb \\
b\text{<C>} &::= bc \\
c\text{<C>} &::= cc
\end{aligned}
$$

The corresponding grammar-based proof requires the strengthening of Induction Hypotheses. For a finite word α consisting of terminal and non-terminal symbols, and for a symbol x (be it terminal or non-terminal), we write $\#(x, \alpha)$ for the number of occurrences of symbol x in word α. We begin our proof with two steps that show that this grammar can only generate words that are in $\mathscr{L}$:

1. **The number of occurrences of a, of b or <B>, and of c or <C> are equal,** i.e.,

$$\#(a, \alpha) = \#(b, \alpha) + \#(\text{<B>}, \alpha) = \#(c, \alpha) + \#(\text{<C>}, \alpha)$$

As long as only the second clause of the first production rule is being applied, it is easy to see that this claim is true: each derivation step introduces exactly one new symbol from the categories $\{a\}$, $\{b, \text{<B>}\}$, and $\{c, \text{<C>}\}$. The other clause of that production rule and the remaining five production rules do not change these respective numbers of occurrences. (This is another example of showing that a property is an invariant, here invariant over derivation steps.)

Note that this is a strengthening of an Induction Hypothesis: we are only interested in proving something for the terminal symbols a, b, and c but need to reason about b and <B>, as well as about c and <C>.

2. **The letters in generated words are sorted with respect to** $a < b < c$**.** As in the previous step, we need to generalize the Induction Hypothesis so that it captures all relevant word forms (all left-hand sides and right-hand sides of all production rules), not just those that contain only terminal symbols. We can inspect the grammar to conclude that all word forms can be written as $w\alpha$, where w is in $\{a, b, c\}^*$ and is sorted, and where α is a word over alphabet $\{\texttt{<B>}, \texttt{<C>}\}$. Therefore, all derived words of terminal symbols only have the claimed sortedness.

The two steps above show that all words of terminal symbols derived by this grammar are contained in $\mathscr{L}$. What about the converse? In other words, is every word in $\mathscr{L}$ derivable from that grammar? Since the elements of $\mathscr{L}$ are parameterized by n in $\mathbb{N}$, we may use Mathematical Induction to show this converse claim. We sketch how to prove this and leave it as an exercise to provide the full details of this proof:

a) It is easy to see that the word form $a^n \texttt{<S>} (\texttt{<B><C>})^n$ may be obtained through the n-fold application of the second clause of the first production rule. Subsequently using the first clause of that production rule to generate an empty word, this gives us an $(n+1)$-step derivation of the word form $a^n (\texttt{<B><C>})^n$.

b) The word form $a^n (\texttt{<B><C>})^n$ from previous item a) may be sorted by means of the second production rule only, such that all symbols $\texttt{<B>}$ occur before all symbols $\texttt{<C>}$. After $\frac{n\cdot(n-1)}{2}$ derivation steps this generates for us the word form $a^n \texttt{<B>}^n \texttt{<C>}^n$.[11]

c) All occurrences of $\texttt{<B>}$ and $\texttt{<C>}$ in the word form $a^n \texttt{<B>}^n \texttt{<C>}^n$ may be converted into terminal symbols. This relies on the remaining four production rules, requires $2n$ derivation steps, and does not change the order in which these symbols or their terminal equivalents occur.

Steps 1 and 2 of the proof above illustrate the Principle of Grammar-Based Induction: the need to define Induction Hypotheses that are tailored for respective left-hand sides of production rules, and the management of the required number of derivation steps. The type of a grammar does not really play a role in how this principle works and so it may be applied for arbitrary grammars of the Chomsky Hierarchy.

If we look at this even more abstractly, we recognize that this principle is merely a generalized inductive proof along the derivation processes and the number of steps that they require. A peculiarity of this principle, however, is that the description of the form of these derivation processes (namely the given grammar itself) is used explicitly in these proofs. In Informatics, we often encounter such explicit use in inductive proofs. For example, there are Induction Principles over the structure of proof systems, proof calculi [5], over the computation sequences or the operational semantics of programs, over the iterations of complex program transformations, and so forth. The only thing that needs to be kept in mind when designing such inductive

---

[11] This step corresponds exactly to the sorting procedure known as *Bubble Sort* in the literature.

proof principles is that the property that is to be proved must be realizable in finitely
many iteration steps.

5.8 Learning Outcomes

The study of this chapter will give you the learning outcomes and abilities discussed
below.

- You will get a first feel for the concept of partial preorders and partial orders and
  their applications.

- You will understand the meaning and significance of the Principle of Well-
  Founded Induction, and of the Descending Chain Condition (Theorem 5.2) and
  its application in proofs of program termination.

- You will be able to do proofs using different kinds of Well-Founded Induction
  principles.

- You will know the particularities of each of the four kinds of Well-Founded Induction principles, including their relative strengths and weaknesses.

- You will have taken in the interplay between inductive definitions of sets (and
  properties) and their corresponding (inductive) proof principles.

- You will be familiar with the expressive power of inductive definitions of sets:

  - What is characteristic about such definitions?

  - What advantages do such definitions offer, especially for Informaticians?

  - Are there alternative forms of definition?

  - Where lie the limits of inductive definitions?

5.9 Exercises

Exercise 5.1 (Divisibility relation and partial orders).

1. Let $|$ be the divisibility relation as defined on page 32. Show that $(\mathbb{N}, |)$ is a partial
   order. Recall that $n|m \Leftrightarrow_{df} \exists k \in \mathbb{N}.\, n \cdot k = m$.
2. Is $(\mathbb{Z}, |)$ with $|$ defined as $x|y \Leftrightarrow_{df} \exists z \in \mathbb{Z}.\, x \cdot z = y$ also a partial order? Justify
   your answer.

Exercise 5.2 (Subset inclusion as partial order).
Let A be a non-empty set. Show that $(A, \subseteq)$ is a partial order.

Exercise 5.3 (Reflexivity, antisymmetry, and transitivity are independent).
A partial order $(A, \sqsubseteq)$ requires that $\sqsubseteq$ be reflexive, antisymmetric, and transitive. Show that these three properties are logically independent of each other, even over finite sets. That is, show that:

1. There is a finite set A and a binary relation $R \subseteq A \times A$ such that R is reflexive, and antisymmetric, but not transitive.
2. There is a finite set A and a binary relation $R \subseteq A \times A$ such that R is antisymmetric and transitive but not reflexive.
3. There is a finite set A and a binary relation $R \subseteq A \times A$ such that R is reflexive and transitive but not antisymmetric.

Exercise 5.4 (Partial order induced by preorder).
Let $(A, \precsim)$ be a preorder over a non-empty set A.

1. Show that $\sim$ defined by $a \sim b$ iff ($a \precsim b$ and $b \precsim a$) is an equivalence relation over the set A.
2. Let $X \leq Y$ be defined by $x \precsim y$ for some x in X and y in Y, where X and Y are equivalence classes of $\sim$ from the previous item. Show that $\leq$ is a partial order over the set of equivalence classes of $\sim$.

Exercise 5.5 (Symmetric preorders).
Let A be a non-empty set and R a homogeneous relation over A that is also symmetric: for all a and b in A we have that (a, b) in R implies (b, a) in R.

1. Let R also be a partial order. Show that, for all a and b in A, we have that $a \leq b$ implies that a equals b.
2. Find instances of A and R such that R is a preorder but not a partial order, and where there are $a \neq b$ with $a \leq b$.
3. Show that all preorders that are also symmetric are equivalence relations.
4. Show that for each non-empty set A, there is exactly one partial order over A that is also symmetric.

Exercise 5.6. Right cancellation of addition
Use the Peano Axioms to prove that addition over the natural numbers allows for *right cancellation*: for all x, y, and n in $\mathbb{N}$ we have that $x + n = y + n$ implies that $x = y$.

Exercise 5.7 (Preorder of implication over set of Boolean terms).
Show that the relation $\precsim$ defined over Boolean terms on page 165, where $t_1 \precsim t_2$ iff $t_1 \Rightarrow t_2$ is logically valid, is a preorder.

Exercise 5.8 (Dense orders).
Let R be a homogeneous binary relation over a non-empty set A. We call R *dense* if for all a and b in A, we have that (a, b) in R implies that there is some c in A with (a, c) in R and (c, b) in R.

1. Let R above be reflexive. Show that R is dense.
2. Let R be irreflexive, i.e., for all a in A we have $(a,a) \notin R$. Let R be dense. Show that A cannot be finite.
3. Show that $(\mathbb{Q}, <)$ is a dense order, where $<$ is the strictly-less-than relation over rational numbers.

Exercise 5.9 (Number of minimal, maximal elements for preorder).

For each scenario below, find a suitable preorder $\sqsubseteq \subseteq A \times A$ and a subset B of A such that

1. B has neither minimal nor maximal elements
2. B has exactly one minimal and exactly one maximal element
3. B has one maximal element and two minimal elements
4. B has three minimal elements and countably infinitely many maximal elements.

Exercise 5.10 (Well-Founded Induction).

1. Show that the divisibility relation on the natural numbers (see Exercise 5.1.1) is a well-founded partial order.
2. Use Well-Founded Induction on the divisibility relation to prove the following statement: *"For every natural number $n \geq 1$ there are natural numbers k and m such that $n = 2^k \cdot m$ and m is odd."*

Exercise 5.11 (Well-Founded Induction: commutativity of addition).

Provide the proof details for Case 2 of the proof on page 173.

Exercise 5.12 (Well-founded peorder).

Show that the relation defined in (5.9) on page 175 is a well-founded preorder. Is it also a partial or total order?

Exercise 5.13 (Lexicographical well-founded preorder).

Let $<$ be the strict and $\leq$ the non-strict usual order over the set of natural numbers $\mathbb{N}$. For the relation $\leq_{lex}$, defined in (5.10) on page 176, show:

1. $\leq_{lex}$ is a preorder
2. $\leq_{lex}$ is well-founded. [Hint: Use Proof By Contradiction and analyze an infinite strictly descending chain.]

Exercise 5.14 (Induction proof).

Prove that every natural number n satisfying $n \geq 8$ can be written as the sum of zero or more copies of the numbers 3 and 5. (For example, for $n = 14$ we have $14 = 3 + 3 + 3 + 5$.)

Exercise 5.15 (Structural Induction).

Prove that all Boolean terms that do not contain any variables are semantically equivalent to either T or F. Hint: Use induction over the structure of the term t.

Exercise 5.16 (Algorithm derived from Structural Induction proof).
Recall the proof by Structural Induction for Theorem 5.4 on page 181.

1. State a recursive definition for a function $AND\_NOT\_ONLY(\phi)$ that computes, for each formula ϕ, a desired formula ϕ' as claimed in that theorem.
2. Convince yourself that the proof by Structural Induction of Theorem 5.4 can, essentially, be reused to proof the correctness of your recursive definition for function $AND\_NOT\_ONLY(\phi)$.

Exercise 5.17 (Functional completeness: same set of atoms).
Consider the following variant of Theorem 5.4 on page 181: for all formulas of propositional logic ϕ, there is a formula of propositional logic ϕ' such that

1. ϕ' and ϕ are semantically equivalent,
2. ϕ' and ϕ contain the same set of atomic propositions, and
3. each logical operator in ϕ' is from the set $\{\neg, \wedge\}$.

Use Structural Induction to prove this variant of the theorem.

Exercise 5.18 (Structural Induction as Well-Founded Induction).
Recall the Proof Principle 12 of Structural Induction on page 180 and the well-founded partial order $\sqsubseteq$ defined in (5.13) on page 180 for an inductively defined set M.

Argue that use of Structural Induction for the set M and a proposition $\mathscr{A}$ is essentially the same as use of Well-Founded Induction over $\sqsubseteq$ for that set M and proposition $\mathscr{A}$.

Exercise 5.19 (Validity of Proof Principle of Minimal Counterexample).
Prove Theorem 5.3, i.e., that the Proof Principle of Minimal Counterexample is valid.

Exercise 5.20 (Induction and Fibonacci numbers).
Recall the introduction of the Fibonacci numbers in Definition 5.6 on page 178. Use an appropriate induction principle to prove the following proposition for all n in $\mathbb{N}$:
$$\sum_{i=0}^{n} (fib(i))^2 = fib(n) \cdot fib(n+1)$$

Exercise 5.21 (Mathematical Induction may be cumbersome).
Prove (5.12) on page 178 using Mathematical Induction. Then compare your proof to the one using Course of Values Induction on that same page.

Exercise 5.22 (Mathematical Induction).
Prove the Pigeonhole Principle 7 on page 80 in both variants.

Exercise 5.23 (Inductive proofs).
Recall your solution to Exercise 4.5 on page 157: an inductive definition of the number of k-element partitions of a set of n elements. Use an appropriate Induction Principle to prove the correctness of your answer to Exercise 4.5.

Exercise 5.24 (Mathematical Induction).
Recall Theorem 5.5 on page 183, for which we already proved the properties *Associativity* and *Commutativity*. Use Mathematical Induction to prove the remaining properties:

1. *Commutativity* of multiplication
2. *Neutral Elements*
3. *Right Cancellation Rules*
4. *Distributivity*

Exercise 5.25 (Examples of Mathematical Induction).
In Example 5.5 on page 186 we proved the claim of its first item by Mathematical Induction already. Now use Mathematical Induction to prove the claims made in the remaining items.

Exercise 5.26 (Negation and negation normal form).
Let ϕ be a formula of propositional logic equal to $\neg\psi$ for some formula ψ of propositional logic that is not an atomic proposition. Show that ϕ is not in negation normal form.

Exercise 5.27 (Recursive definition of NNF conversion).
Recall the discussion in Example 5.6 on page 188, which highlighted the need to strengthen the Induction Claim to prove that each formula of propositional logic has a formula in NNF that is semantically equivalent to it.

1. Specify a recursive definition $ToNNF(\phi)$ that converts arbitrary formulas of propositional logic (defined over atoms with operators $\neg$, $\wedge$, and $\vee$ only) into semantically equivalent formulas that are in NNF.
2. Use Well-Founded Induction to show that your recursive definition, for all input formulas ϕ:

   a. terminates
   b. outputs a formula that is semantically equivalent to ϕ, and
   c. outputs a formula that is in NNF.

Exercise 5.28 (Proof of Substitution Lemma).
Complete the proof of Lemma 5.2 on page 190, by proving its Case 2 and Case 7.

Exercise 5.29 (Proof rule for assignment in Hoare Calculus).
Recall the discussion of proof rule *ASS* on page 193. Show that this rule allows us to prove that $\{x*y > y\}\, x = x*y\, \{x > y\}$ holds, i.e., whenever $x*y$ is larger than y, then the assignment $x = x*y$ leads us to a state in which x is larger than y.

Exercise 5.30 (Proof rule *CONS* and proofs of logical consequence with theories).
Recall the discussion on page 196, where

$$(x = x_0 \wedge y = y_0) \Rightarrow ((x+y) - ((x+y) - y) = y_0 \wedge ((x+y) - y) = x_0) \quad (5.28)$$

was the only non-trivial logical consequence of a proof using rule *CONS*.

Use the familiar axioms of arithmetic to show the logical consequence in (5.28), i.e., that this formula is valid (always true).

Exercise 5.31 (Showing that all words of a language are generated by a grammar).
Recall the proof sketch on page 202 that all words of the language $\mathscr{L}$ in (5.27) are generated by a monotone grammar.

Provide the full details of that proof, based on this proof sketch. In particular, make clear how Mathematical Induction is being used in that proof.

Exercise 5.32 (Tic-tac-toe game: reachable configurations).
Recall the game of Tic-tac-toe from Section 3.4.1 on page 99.

1. Use the Proof Principle of Minimal Counterexample to show that no configuration in the set $\mathsf{Conf}_X \cap \mathsf{Conf}_O$ can be reached in any play of that game. [Hint: Define a suitable well-founded order on the set of all plays.]
2. Given any configuration (and no other information), define a predicate that is true iff that configuration is reachable in some play of that game.

Chapter 6
Inductive Approach:
Potential, Limitations, and Pragmatics

God made the integers. Everything else is the work of man.
(Leopold Kronecker)

In the preceding chapters we saw a definite *Best Practice* of Informatics: the consistently executed inductive approach, leading from the inductive definition of the objects of study and the formulation of properties we wish to prove about such objects to the concluding proofs that all such objects have these properties. Any deviation from this best practice should be carefully justified. In these concluding sections, we want to discuss the potential as well as the limitations of this best practice of inductive approaches from the point of view of Informatics and its corresponding pragmatics.

6.1 Potential

A prominent example of the inductive approach is the definition of syntax and semantics of term languages and languages of formulas, as we illustrated in Sections 4.2.1, 4.3.3, and 4.4 through the consideration of Boolean terms. The machinery we built to that end was constructed in a completely inductive and so systematic way. This not only provided formal rigor and unambiguous formulations, it also meant that even core results such as the Substitution Lemma could be proved by structural induction with relative ease. The reason for the simplicity of proofs for such fundamental results is the coordinated manner in which different inductive definitions are given so that they guarantee *compositional reasoning*: the semantics of to-be-considered properties can be proved stepwise, along the inductive structure of the objects of study. This not only simplifies the structure of proofs. Compositionality is perhaps the most powerful tool in Informatics for achieving modular design and thereby scalability of development and of accompanying reasoning to support correctness, reliability, performance, and security.

We may think of compositional reasoning as a strategy of *divide and conquer*, where a complex problem is divided into several but smaller (and therefore simpler) problems such that the solutions of these subproblems can be easily synthesized into a solution of the overall, composed problem. Moreover, since the inductive

© Springer International Publishing AG, part of Springer Nature 2018
B. Steffen et al., *Mathematical Foundations of Advanced Informatics*,
https://doi.org/10.1007/978-3-319-68397-3_6

approach works alongside the inductively defined structure, the overall effort of solving problems typically grows only in proportion with the size of the inductively defined structure.

This discussion shows that there is a principled incentive for taking an inductive approach. And in cases in which the use of an inductive approach seems impossible or too complicated, it is still advantageous to search for generalizations of inductive approaches so that as many as possible of the usual structures, properties, and methods of these approaches can be adapted successfully. We illustrate this point by the transition from a simple term language, such as the one we studied for Boolean terms, to a more expressive programming language. This historically led to a generalization of inductive approaches to so-called *Denotational Semantics*. The problem at hand here is to extend inductive definitions of semantics (for example, of Boolean terms) to programming constructs such as loops or recursive procedures such that the intended computational process represented by program terms can follow these inductive semantic definitions in the usual stepwise and hierarchical manner.

However, the characteristic feature of loops and recursive procedures is that they are not hierarchical but they evaluate the same program point repeatedly yet in potentially different computational states. In particular, an evaluation step may not necessarily result in what would syntactically be a sub-program. In order to control this phenomenon, Denotational Semantics introduces operations that work on the semantic level and are based on insights from *Fixed-Point Theory* (cf. [20] and the second volume). Such operators can then compute, in a single step, the entire effect of the loop-specific iteration processes, based on the body of the loop code and on the termination condition for that iteration.

In this manner, we arrive at a fixed-point semantics that solves this problem in an elegant way, and this solution has also contributed to the development of functional programming languages and to the mastering of functions of higher order; the latter are functions that may have functions as input or output. However, this success story of fixed-point semantics cannot be repeated in the treatment of parallel computation, for example in the introduction of dynamic process or thread creation.

This is one reason why modern programming and processor languages are increasingly based on so-called *operational* semantics. Such semantics typically capture the effect of a single computation step in a manner that more closely resembles how a program will be executed in a run-time environment. Although operational semantics are not as elegant as denotational ones, for example they typically do not support compositionality, they are easily extensible. This then supports the incremental design and implementation of programming languages: an operational semantics for a core language can be extended in unison with extensions of that core language for new constructs such as procedures, parallel execution, and recursion [67]. We therefore see that there are good reasons for abandoning the best practice of inductive approaches in this case. However, this does not mean that we should give up the best practice of inductively defining the relevant syntax of programming languages and expressions used in the operational semantics!

Gordon Plotkin's approach of presenting operational semantics in a structured way is, in that regard, the most mature one [64]. He apparently tried to preserve inductive thinking as much as possible so that inductive proofs along the corresponding sequences of computation steps no longer look too different from conventional structural induction proofs. But even this structured approach to operational semantics represents a departure from the compositional approach to denotational semantics, since it formulates thinking in terms of the evaluation of computation steps in sequence, rather than in meaning functions that holistically capture entire (sub-) programs.

This departure seems inevitable as modern programming languages comprising, e.g., special forms of parallelization or acceleration, seem no longer to admit an adequate structure-preserving mapping from program representations into semantic domains consisting of compositional meaning functions. The attempts to maintain a denotational semantics, such as the introduction of power domains, essentially failed (at least from a practical perspective). The price to be paid in terms of complexity of the semantic domain seems too high to be competitive in comparison to Plotkin's operational semantics, which tends to seamlessly deal with language extensions. However, this does not mean that compositionality has been abandoned entirely. Rather it has become an art of where to draw the line. Illustrative for this research direction are the *Theory of Distributed Systems* and related areas which aim at the development of foundational concepts for the mastering and control of parallelism, where parallel executions may also have to consider real-time aspects (for example in cyber-physical systems [58]) and stochastic behavior (for example in adaptive software [24]).

6.2 Dealing with Limitations

The development of programming languages, as discussed in the previous section, should have made clear that even *Best Practice* encounters its limits. But this is not a fundamental problem. Rather, it is crucial that we develop an instinct for what best to do when this best practice fails as prescribed. In such a case, should we try to "rescue" that principle, as done in Denotational Semantics through the introduction of fixed-point theory, or should we simply abandon the inductive approach? Put in another way, how much effort should we invest in rescuing the inductive approach for a problem at hand, and when is it better to devise new, alternative methods to solve that problem?

We may illustrate this in the comparison that Robin Milner once made to situations encountered in playing golf, which we already alluded to in the Preface of this book. Consider a player who faces a hazard: her ball sits in a bunker or is immersed in a water hazard such as a pond. When should a player make a spectacular shot at the hole, and when would it be better for her to just take the penalty (for example, one that results from moving the ball to a position at which it can be played better)? It is difficult to make optimal decisions in such situations, even for professional players who have a lot of experience. In Informatics, we may face more complex, abstract systems for which it is even more complicated to make the right decisions.

But for such systems, we may want to follow instead a relatively simple rule for design, implementation, and validation: if our metaphorical ball is in a bunker, we may feel justified to play this ball as usual. But if our metaphorical ball sits at the bottom of a pond, we should take our losses and move on to developing new methods to solve the problem at hand. One of the aims of this book was to develop a better intuition for when the Informatics ball sits on the green, in a bunker, or at the bottom of a pond – so that we may formulate and follow such heuristic rules for choosing appropriate and effective methods.

Let us illustrate this with the termination proof from Section 5.3.1 (page 174). The termination proof for the algorithm that computes the greatest common divisor is clearly a putt that takes place on the green. Every Informatician should be able to complete such proofs. Things are already somewhat different for the termination proof of the Ackermann function. This requires knowledge of well-founded induction and of one of its instances, the lexicographical ordering on $\mathbb{N} \times \mathbb{N}$. We may say that this termination proof is taking place in the bunker: having the right knowledge, we realize that this is a rather straightforward proof by well-founded induction and so the proof becomes rather elegant and we get the ball back onto the green.

Things are very different indeed when we consider the Collatz function. Upon first inspection, its definitional body seems even easier than the function for the greatest common divisor. However, we cannot recognize any well-founded ordering for this function that would show, without fail, that every execution of this function – which either halves an even input or adds 1 to the result of multiplying an odd input by *three* – strictly decreases the computational state in that well-founded ordering. The ball for this termination problem is therefore very far from the green, at the bottom of a perhaps bottomless pond – meaning that we may not have the mathematical tools to ever prove this claim.

But it is not clear whether the ball for the Collatz problem will stay in such a pond. History has shown that there are problems whose balls are in a pond but that, upon closer and longer inspection, end up being in a bunker or even on the green. For example, it was a long-standing open problem whether deciding that a natural number n is prime can be done efficiently (in computation time that is polynomial in the size of the input, the binary representation of n). When this was proved in the positive, the proof did not really have to invent radically new methods; it "just" had to combine existing methods in an ingenious manner [1].

Things turned out to be quite different for an, at first sight innocent-looking question: are *regular languages* closed under complementation [35]? Recall that a formal language L is a subset of A^* for some alphabet A. Regular languages, a particular subclass of formal languages of high practical importance, can be described via *regular expressions* whose syntax is defined inductively by the following BNF:

$$L ::= \emptyset \mid \{a\} \mid L \cdot L \mid L + L \mid L^*$$

In this definition a can be any element of alphabet A. Semantically, regular expressions denote the smallest set of words that contains $A \cup \{\varepsilon\}$ and are closed under composition (for ·), union (for +), and the Kleene *, i.e., finite iteration of concatenation.

Just looking at the inductive definition of regular expressions does not seem to give one any route to approach this problem. It required Kleene's fundamental theorem [35], to prove that regular languages are closed under complementation, i.e., that, given regular L, $A^* \setminus L$ is also a regular language.

Kleene's theorem establishes an, in retrospect rather natural correspondence between regular expressions and finite automata, and thereby an alternative language representation that is adequate to answer the question. Using the imposed straightforward back and forth translation, the complementation problem for regular languages can be reduced to a complementation problem for finite automata: given a finite automaton $\mathscr{A}_L$ over alphabet A whose language is $L \subseteq A^*$, find a finite automaton $\overline{\mathscr{A}_L}$ whose language is the complement of L, hence $A^* \setminus L$.

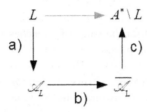

Fig. 6.1 Complementation of regular languages via (b) automata complementation. While arrow a) represents straightforward inductive translation of regular expressions into nondeterministic automata followed by determinization and minimization, arrow c) refers to the construction according to Kleene's theorem

Automata complementation is a standard problem in every theory course. Its solution hinges on two clean operations on automata: *determinization* and swapping accepting and rejecting states. These steps are comparatively straightforward, although determinization is potentially computationally expensive and may lead to an exponential blowup of the size of the automaton. Consequently, the regular expression resulting from complementation may grow exponentially, which gives a strong hint that a compositional complement operation at the level of regular expressions is indeed infeasible.

In contrast to the termination of the Collatz function, complementation of regular expressions has found a conceptually surprisingly elegant solution, even though, in our golf analogy, it resides at the bottom of a pond concerning compositionality: determinizing a finite automaton is a truly non-compositional activity. Algorithmically, the imposed potential exponential blowup in size often constitutes a bottleneck, similar to the one when moving to power domains in order to compositionally deal with parallel systems. Dealing with such bottlenecks is a major research direction in areas such as program and system verification [5, 25].

We have frequently encountered the impact of change of representation in our own research, for example in the context of program analysis. Traditionally, each of the numerous analysis problems, such as dead code detection, liveness of variables, and reaching definitions, have been hand coded one by one. Dataflow analysis frameworks improved on this situation by supporting specifications in terms of (Boolean) equation systems whose minimal or maximal solutions determined the desired properties. Compared to the traditional approach, equation systems raised the level of abstraction and freed the analysis designer from tedious programming work. In a sense they reduced the development effort by no longer requiring a detailed description of HOW the analysis has to be performed but only a WHAT-style description of the analysis goal to be achieved. The underlying change of representation and reasoning drastically improved the development performance but required the developer to think in terms of the implied fixpoint computation and not just about the envisioned analysis problem. For example, liveness of a variable v at some program point p does not simply mean that its value may still be required for subsequent computation, but is considered as a backward analysis problem which determines whether a request for the value of v reaches p. This means that one has to understand the fixpoint computation, which itself is a HOW, in order to understand what the equation system means. The situation significantly improves when one specifies the program analysis in terms of temporal logic properties. The property of liveness of a variable or, as one could say, the "true" WHAT specification becomes

there is a path starting at p on which v is used before its value is modified or the program terminates

which is a simple *unless* property in temporal logic [68, 69].

This gain in abstraction may not seem very impressive. However, it is crucial when it comes to verifying properties of the program analyses. This impact became apparent during our construction of the lazy code motion algorithm [42, 43]. The possibility to refine the WHAT specification by conjoining it with other WHAT specifications, which is typically impossible for HOW specifications, let us elegantly and efficiently solve a 15-year-old problem (see, e.g., [22]) in dataflow analysis. In fact, the corresponding temporal specification runs faster on a classical iterative model checker [70] than the weaker original handwritten algorithm.

Key to this solution was the change of representation: temporal logic formulas, like any logical specification formalism, support conjunction and disjunction, and thereby allow one to simply compose the desired complex specification as a logical combination of simple properties. In this case, moving to a logical specification formalism not only led to a straightforward (executable) problem specification but also to elegant correctness and optimality proofs. The impact of this change becomes particularly apparent when compared to the argumentation required in the seminal paper on code motion [48].

And there was another benefit: the algorithms specified in terms of temporal formulas worked directly also for an interprocedural setting when using a model checker for context-free systems [11, 12, 70], thus demonstrating the superiority of WHAT descriptions when it comes to adaptation and migration.

After this discussion, it may not be too surprising that the mentioned model checker for context-free systems is, essentially, also based on a change of representation: the move from a first-order specification in terms of properties to a second-order specification in terms of transformation of properties (i.e., in terms of a functional representation) allows one to compositionally deal with the structure of context-free systems.

We should hasten to add that breakthroughs like Kleene's theorem, which can be regarded as a prime example of exploiting the freedom of representation, or even better, the power of changing representation, are quite rare and achieving them is not part of the daily routine of an Informatician. An Informatician, as a professional, should therefore not be concerned with the resolution of problems such as the termination of the Collatz function. Rather, she should pay attention that her design, implementation, validation, change management, and retirement decisions always keep the system ball on the green, where we do not encounter problems of the Collatz variety. And even if we did, she would seek methods that can bring the system back onto the green, even if that would incur calculated risks and penalties. We mean to emphasize that such an overall aim requires a sharp awareness of problems that do or may arise, whose management can only be founded on appropriate mathematical techniques such as the ones featured in this trilogy.

6.3 Pragmatics

As discussed in this book, the most powerful proof principles of induction are Proof Principle 9 of Well-Founded Induction and Proof Principle 10 of Minimal Counterexample. The latter we applied, for example, to show the correctness of the Well-Founded Induction Principle. The Proof Principle of Minimal Counterexample is often used in Mathematics, since such use ensures that the failure to complete an induction proof was not due to choosing a too weak induction principle, where by "too weak" we really mean "less effective in this problem instance": two methods may be equally powerful mathematically but one may be easier for a human or machine to invoke on a given problem. In Group Theory, for example, there are proofs that cover many pages in order to derive a minimal counterexample to a claim one wishes to prove, only to then derive inconsistent propositions which therefore refute the existence of a minimal and so of *any* counterexample to the claim. A prominent example here is the hundreds-of-pages-long proof of the Odd-Order Theorem which essentially follows this structure.

But we also saw that some problems, such as the Whitehead problem for groups with commutative group multiplication, cannot be solved within a standard system of set theory. So the very powerful method of any mathematical proof within such a set theory is not powerful enough to solve that problem. Again, we may react to this by either seeking alternative forms of set theory in which this can be proved (but then we have the possibility that alternative extensions of set theory may prove the opposite fact!), or we simply interpret this as advice to file that problem as being insoluble.

To return to the different Induction Proof Principles covered in this book, you may ask why there is such a range of principles. The answer is that specific problems benefit from specific proof processes, and that such principles can capture such needed specifics. The Principle of Structural Induction, for example, was seen to specify the structure of induction proofs in quite some detail. And such level of detail in form is a great advantage when it comes to automating such proofs in computer programs: at the very least we may completely automate the required case analysis and the proof obligations for each required induction case.

For inductive proofs over BNFs, or more generally over grammars, we may use the left-hand sides of grammar clauses or rules to appropriately structure an Induction Hypothesis for this particular case. Of course, we may group all these hypotheses into a global one – as we indicated in Proof Obligation 1 on page 198. But such a more general Induction Hypothesis does not aid human comprehension and just adds complexity to managing the corresponding induction proof. We leave it as an exercise to revisit the Induction Proof for the monotone grammar that generates the language in (5.27) on page 201.

Most of the Induction Principles that we introduced in this book may therefore be seen as specific, parameterized proof patterns which the human user or automated prover can instantiate or simplify in order to guide and complete the inductive proof. In contrast, the Proof Principle of Minimal Counterexample does not offer any guiding support. The prover – be it a human being, a machine or both interacting – therefore needs to draw from other methods to successfully apply this proof principle.

The development of appropriate and effective patterns – for proofs, data structures, software architectures, processes, and so forth – is a central task in Informatics. Such developments as, for example, the completion and documentation of a proof for a design can then be made accessible to a wider user base with varying levels of expertise. This use of concepts in a simpler way, and their reuse with relative ease, is also pursued in other parts of Informatics, for example in the inheritance principle in object-oriented programming, and in the creation and use of software libraries in application programming. These concepts and methods are all devised so that their use avoids redundant work, and so that completed artifacts can – in principle – be made available to other potential users. The open-source code movement, say the development of the Linux kernel, is a good example of how such sharing of work amplifies the adoption, impact, and reliability of written code bases.

A promising concrete approach in this context is the so-called *domain-specific languages* (DSL) and their corresponding *frameworks*. In the extreme case, such frameworks can be used by domain-specific problem owners who do not have any technical knowledge of programming; for example, the DSL may represent programs as graphs of domain objects that users can connect in a way that is meaningful to domain experts [53]. Although such user-facing DSLs have appeal, it is typically non-trivial to write a software system in which a user-facing DSL has the right level of abstraction for users to make competent design and implementation decisions, and to then transform such user-facing programs written in that DSL into syntax that can be operationalized in the domain itself, which may be an embedded system with a CPU on it, an entire industrial control system, the enterprise software of an online grocery company, and so forth [52]. One challenge here is that users think and express concepts at their semantic level of abstraction and from a WHAT perspective (for example, *"deliver goods to delivery address once goods are loaded into van"*), whereas operational realizations operate at another, typically lower, level of semantic abstraction which expresses the corresponding HOW (for example, *"the hatch of the van has to be closed, the driver must then push the ready-to-go button and make sure that all sensors report normal safety before she is allowed to start the engine, and follow the directions of the navigation system, ... "*) [46]. This requires that the transformation of more abstract representations (which tend to have a more declarative character) into more concrete and therefore more operational ones does not introduce inconsistencies that would corrupt the intended functioning of the system.

We may also interpret the rigorous inductive approaches developed in this book as a foundation of the so-called *Design for X* paradigm, in which we explicitly conceptualize scenarios that allow us to express aims and techniques for realizing them. One prominent instance of this paradigm from the realm of the hardware industry is where X is testability, the *Design for Testability*. This approach advocates the design of hardware circuits in such a way that it allows for the extension of hardware interfaces, through which specific types of errors can be detected and localized. The Design for X paradigm is conceptually related to methods by which we strengthen Induction Hypotheses: what do you need to know in order to draw the necessary conclusions for an inductive step? In practice, it is important to express or represent this *What* in a manner that allows for efficient if not completely automatable reasoning in inductive proofs. First-order logic (including its variants) has established itself as an effective means of specifying this *What*. But efficient automated reasoning in first-order logic is only possible for specific, limited proof patterns. Automated provers therefore need to identify such patterns and then apply them with rigor and formality.

The second volume of this trilogy, *Algebraic Thinking*, focuses on another crucial instrument for perfect modeling in Informatics: *abstraction*. The essential ingredient of a good abstraction is that it still captures all the salient properties of the scenario it intends to model; where the determination of which properties are salient is a function of the properties we mean to prove about the intended system. Therefore, a good abstraction typically needs to reflect well the functionalities, operations, and structure that pertain to reasoning about a system. One success story of such a good abstraction can be seen in *Types in Programming* and type inference. As a simple example, we may give a program expression a type, for example that it is a function that maps two integers to a Boolean. This can be done by a declaration through a programmer, or it may be automatically inferred from a program that has no type annotations. Such a type stipulates a *run-time invariant*: whenever we supply two integers to this function at run-time, then we can rely on it to compute a Boolean since this is its declared type. Types can make run-time systems much more reliable and more efficient, and they are of great help in revealing oversights. Of course, this only works when the type annotations are semantically correct, which is by no means trivial for complex programming languages.

Abstract algebra provides a very mature mathematical framework in which we may formulate and engineer abstractions through concepts such as homomorphism (i.e., the preservation of structure), sub-structures (e.g., subspaces in a vector space), and quotient structures (e.g., the natural numbers modulo a prime number). This framework allows us to recognize the tension between the demands to preserve all structure and to abstract more aggressively – for example, to support more scalable algorithms. But this framework also often allows us to identify the obstacles to effective abstraction, and it offers techniques for dealing with or circumventing such obstacles. Inductive approaches and algebraic thinking will then be combined in the third volume of this trilogy in order to illustrate the art of perfect modeling.

References

1. Manindra Agrawal, Neeraj Kayal, and Nitin Saxena. PRIMES is in P. *Annals of Mathematics*, 160(2):781–793, 2004.
2. Alfred V. Aho, Monica S. Lam, Ravi Sethi, and Jeffrey D. Ullman. *Compilers: Principles, Techniques, and Tools (2nd Edition)*. Addison-Wesley Longman Publishing, Boston, MA, USA, 2006.
3. Dana Angluin. Learning regular sets from queries and counterexamples. *Inf. Comput.*, 75(2):87–106, 1987.
4. Andrew W. Appel and Jens Palsberg. *Modern Compiler Implementation in Java*. Cambridge University Press, New York, NY, USA, 2nd edition, 2003.
5. K. R. Apt, F. S. de Boer, and E.-R. Olderog. *Verification of Sequential and Concurrent Programs, 3rd Edition*. Texts in Computer Science. Springer-Verlag, 2009. 502 pp, ISBN 978-1-84882-744-8.
6. Adnan Aziz, Amit Prakash, and Tsung-Hsien Lee. *Elements of Programming Interviews: 300 Questions and Solutions*. CreateSpace Independent Publishing Platform, USA, 1st edition, 2012.
7. Clark Barrett, Roberto Sebastiani, Sanjit A Seshia, and Cesare Tinelli. Satisfiability modulo theories. *Handbook of Satisfiability*, 4, 2009.
8. Bernhard Beckert, Reiner Hähnle, and Peter H. Schmitt. *Verification of object-oriented software: The KeY approach*. Springer, 2007.
9. Ella Bounimova, Patrice Godefroid, and David A. Molnar. Billions and billions of constraints: whitebox fuzz testing in production. In *35th International Conference on Software Engineering, ICSE '13, San Francisco, CA, USA, May 18-26, 2013*, pages 122–131, 2013.
10. Alan Bundy, Robert S. Boyer, Deepak Kapur, and Christoph Walther. Automation of proof by mathematical induction. Report of the Dagstuhl seminar 30/95, 1995. Available via http://www.dagstuhl.de/Reports/95/9530.pdf.
11. Olaf Burkart and Bernhard Steffen. Model checking for context-free processes. In *CONCUR '92, Third International Conference on Concurrency Theory, Stony Brook, NY, USA, August 24-27, 1992, Proceedings*, pages 123–137, 1992.
12. Olaf Burkart and Bernhard Steffen. Model checking the full modal mu-calculus for infinite sequential processes. *Theor. Comput. Sci.*, 221(1-2):251–270, 1999.
13. Cristiano Calcagno, Dino Distefano, Peter W. O'Hearn, and Hongseok Yang. Compositional shape analysis by means of bi-abduction. *J. ACM*, 58(6):26, 2011.
14. Byron Cook, Andreas Podelski, and Andrey Rybalchenko. Terminator: Beyond safety. In *Computer Aided Verification, 18th International Conference, CAV 2006, Seattle, WA, USA, August 17-20, 2006, Proceedings*, pages 415–418, 2006.
15. Byron Cook, Andreas Podelski, and Andrey Rybalchenko. Proving program termination. *Commun. ACM*, 54(5):88–98, 2011.

© Springer International Publishing AG, part of Springer Nature 2018

B. Steffen et al., *Mathematical Foundations of Advanced Informatics*,

https://doi.org/10.1007/978-3-319-68397-3

16. Tom Copeland. *Generating Parsers with JavaCC*. Centennial Books, Alexandria, VA., 2nd edition, 2009.

17. T. H. Cormen, C. E. Leiserson, R. L. Rivest, and C. Stein. *Introduction to Algorithms*. The MIT Press, 3rd edition, 2009.

18. P. Cousot and R. Cousot. Abstract interpretation: a unified lattice model for static analysis of programs by construction or approximation of fixpoints. In *Conference Record of the Fourth Annual ACM SIGPLAN-SIGACT Symposium on Principles of Programming Languages*, pages 238–252, Los Angeles, California, 1977. ACM Press, New York, NY.

19. D.R. Dams. *Abstract Interpretation and Partition Refinement for Model Checking*. PhD thesis, Eindhoven University of Technology, 1996.

20. B.A. Davey and H.A. Priestley. *Introduction to Lattices and Order*. Cambridge mathematical text books. Cambridge University Press, 2002.

21. Leonardo Mendonça de Moura and Nikolaj Bjørner. Satisfiability modulo theories: introduction and applications. *Commun. ACM*, 54(9):69–77, 2011.

22. D. M. Dhamdhere. A fast algorithm for code movement optimisation. *SIGPLAN Not.*, 23(10):172–180, October 1988.

23. E. Allen Emerson. Temporal and modal logic. In *Handbook of Theoretical Computer Science*, pages 995–1072. Elsevier, 1995.

24. Antonio Filieri, Giordano Tamburrelli, and Carlo Ghezzi. Supporting self-adaptation via quantitative verification and sensitivity analysis at run time. *IEEE Trans. Software Eng.*, 42(1):75–99, 2016.

25. Nissim Francez. *Program verification*. International computer science series. Addison-Wesley, 1992.

26. R. Gentilini, C. Piazza, and A. Policriti. From bisimulation to simulation: Coarsest partition problems. *Journal of Automated Reasoning*, 31(1):73–103, 2003.

27. Lou Goble, editor. *The Blackwell Guide to Philosophical Logic*. Wiley-Blackwell, 2001.

28. Carla P. Gomes, Henry Kautz, Ashish Sabharwal, and Bart Selman. Chapter 2 - Satisfiability Solvers. In Frank van Harmelen, Vladimir Lifschitz, and Bruce Porter, editors, *Handbook of Knowledge Representation*, volume 3 of *Foundations of Artificial Intelligence*, pages 89 – 134. Elsevier, 2008.

29. Markus H. Gross. *Visual computing - the integration of computer graphics, visual perception and imaging*. Computer graphics: systems and applications. Springer, 1994.

30. Thomas C. Hales. A proof of the Kepler Conjecture. *Annals of Mathematics (2)*, 162(3):1065–1185, 2005.

31. E. Harzheim. *Ordered Sets*. Advances in Mathematics - Kluwer Academic Publishers. Springer, 2005.

32. Wilfrid Hodges. Classical logic I: First order logic. In Lou Goble, editor, *The Blackwell Guide to Philosophical Logic*. Blackwell Publishers, 2001.

33. Douglas R. Hofstadter. *Gödel, Escher, Bach: An Eternal Golden Braid*. Basic Books, New York, NY, USA, 1979.

34. John Hopcroft. An $n \log n$ algorithm for minimizing states in a finite automaton. In *Theory of machines and computations (Proc. Internat. Sympos., Technion, Haifa, 1971)*, pages 189–196. Academic Press, New York, 1971.

35. John E. Hopcroft, Rajeev Motwani, and Jeffrey D. Ullman. *Introduction to Automata Theory, Languages, and Computation – International Edition (2nd ed)*. Addison-Wesley, 2003.

36. Falk Howar, Bernhard Steffen, and Maik Merten. Automata learning with automated alphabet abstraction refinement. In *International Workshop on Verification, Model Checking, and Abstract Interpretation*, pages 263–277. Springer, 2011.

37. Michael Huth and Mark Dermot Ryan. *Logic in computer science - modelling and reasoning about systems (2nd ed.)*. Cambridge University Press, 2004.

38. Graham Hutton. *Programming in Haskell*. Cambridge University Press, August 2016.

39. Malte Isberner, Falk Howar, and Bernhard Steffen. Learning register automata: from languages to program structures. *Machine Learning*, 96(1-2):65–98, 2014.

40. Malte Isberner, Falk Howar, and Bernhard Steffen. The TTT algorithm: A redundancy-free approach to active automata learning. In Borzoo Bonakdarpour and Scott A. Smolka, editors, *Runtime Verification - 5th International Conference, RV 2014, Toronto, ON, Canada, September 22-25, 2014. Proceedings*, volume 8734 of *Lecture Notes in Computer Science*, pages 307–322. Springer, 2014.

41. Steven C. Johnson. Yacc: Yet another compiler compiler. In *UNIX Programmer's Manual*, volume 2, pages 353–387. Holt, Rinehart, and Winston, New York, NY, USA, 1979.

42. Jens Knoop, Oliver Rüthing, and Bernhard Steffen. Lazy code motion. In *Proceedings of the ACM SIGPLAN 1992 Conference on Programming Language Design and Implementation*, PLDI '92, pages 224–234, New York, NY, USA, 1992. ACM.

43. Jens Knoop, Oliver Rüthing, and Bernhard Steffen. Optimal code motion: Theory and practice. *ACM Trans. Program. Lang. Syst.*, 16(4):1117–1155, July 1994.

44. Tiziana Margaria and Bernhard Steffen. Simplicity as a driver for agile innovation. *IEEE Computer*, 43(6):90–92, 2010.

45. Peter W. Markstein. The new IEEE-754 standard for floating point arithmetic. In Annie A. M. Cuyt, Walter Krämer, Wolfram Luther, and Peter W. Markstein, editors, *Numerical Validation in Current Hardware Architectures, 6.1. - 11.1.2008*, volume 08021 of *Dagstuhl Seminar Proceedings*. Internationales Begegnungs- und Forschungszentrum für Informatik (IBFI), Schloss Dagstuhl, Germany, 2008.

46. Bertrand Meyer and Jim Woodcock, editors. *Verified Software: Theories, Tools, Experiments, First IFIP TC 2/WG 2.3 Conference, VSTTE 2005, Zurich, Switzerland, October 10-13, 2005, Revised Selected Papers and Discussions*, volume 4171 of *Lecture Notes in Computer Science*. Springer, 2008.

47. G.H. Moore. *Zermelo's axiom of choice: its origins, development, and influence*. Studies in the history of mathematics and physical sciences. Springer-Verlag, 1982.

48. E. Morel and C. Renvoise. Global optimization by suppression of partial redundancies. *Commun. ACM*, 22(2):96–103, February 1979.

49. Steven S. Muchnick. *Advanced Compiler Design and Implementation*. Morgan Kaufmann, 1997.

50. Markus Müller-Olm and Oliver Rüthing. On the complexity of constant propagation. In *Programming Languages and Systems, 10th European Symposium on Programming, ESOP 2001 Held as Part of the Joint European Conferences on Theory and Practice of Software, ETAPS 2001, Genova, Italy, April 2-6, 2001, Proceedings*, volume 2028 of *Lecture Notes in Computer Science*, pages 190–205. Springer, 2001.

51. Markus Müller-Olm, David A. Schmidt, and Bernhard Steffen. Model-checking: A tutorial introduction. In Agostino Cortesi and Gilberto Filé, editors, *SAS*, volume 1694 of *Lecture Notes in Computer Science*, pages 330–354. Springer, 1999.

52. Stefan Naujokat, Michael Lybecait, Dawid Kopetzki, and Bernhard Steffen. CINCO: A Simplicity-Driven Approach to Full Generation of Domain-Specific Graphical Modeling Tools. *Software Tools for Technology Transfer*, 2017.

53. Stefan Naujokat, Johannes Neubauer, Tiziana Margaria, and Bernhard Steffen. Meta-level reuse for mastering domain specialization. In *International Symposium on Leveraging Applications of Formal Methods*, pages 218–237. Springer International Publishing, 2016.

54. Anil Nerode and Richard A. Shore. *Logic for Applications (2nd ed.)*. Graduate Texts in Computer Science. Springer-Verlag, 1997.

55. Flemming Nielson, Hanne R. Nielson, and Chris Hankin. *Principles of Program Analysis*. Springer-Verlag, New York, 1999.

56. Hanne Riis Nielson and Flemming Nielson. *Semantics With Applications: A Formal Introduction*. Wiley, New York, USA, 1992. Free preprint available via www.daimi.au.dk/ ~bra8130/Wiley_book/wiley.html.

57. Tobias Nipkow, Lawrence C. Paulson, and Markus Wenzel. *Isabelle/HOL — A Proof Assistant for Higher-Order Logic*, volume 2283 of *LNCS*. Springer, 2002.

58. Cameron Nowzari and Jorge Cortés. Team-triggered coordination for real-time control of networked cyber-physical systems. *IEEE Trans. Automat. Contr.*, 61(1):34–47, 2016.

59. Martin Odersky, Lex Spoon, and Bill Venners. *Programming in Scala, 2nd Edition: A comprehensive step-by-step guide*. Artima Inc, 2011.
60. Peter O'Hearn. From categorical logic to Facebook engineering. In *Logic in Computer Science (LICS), 2015 30th Annual ACM/IEEE Symposium on*, pages 17–20. IEEE, 2015.
61. Gerard O'Regan. *Guide to Discrete Mathematics: An Accessible Introduction to the History, Theory, Logic and Applications*. Springer, 2016.
62. Robert Paige and Robert E. Tarjan. Three partition refinement algorithms. *SIAM J. Comput.*, 16(6):973–989, 1987.
63. Terence Parr, Sam Harwell, and Kathleen Fisher. Adaptive LL(*) parsing: The power of dynamic analysis. In *Proceedings of the 2014 ACM International Conference on Object Oriented Programming Systems Languages & Applications*, OOPSLA '14, pages 579–598, New York, NY, USA, 2014. ACM.
64. Gordon D. Plotkin. A Structural Approach to Operational Semantics. Technical report, DAIMI, Aarhus University, Denmark, 1981.
65. R. L. Rivest, A. Shamir, and L. Adleman. A method for obtaining digital signatures and public-key cryptosystems. *Commun. ACM*, 21(2):120–126, February 1978.
66. David A. Schmidt. *Denotational Semantics: A Methodology for Language Development*. William C. Brown Publishers, Dubuque, IA, USA, 1986.
67. David A. Schmidt. *The structure of typed programming languages*. Foundations of computing series. MIT Press, 1994.
68. Bernhard Steffen. Data flow analysis as model checking. In *Proceedings of the International Conference on Theoretical Aspects of Computer Software*, TACS '91, pages 346–365, London, UK, 1991. Springer.
69. Bernhard Steffen. Generating data flow analysis algorithms from modal specifications. *Science of Computer Programming*, 21(2):115 – 139, 1993.
70. Bernhard Steffen, Andreas Claßen, Marion Klein, Jens Knoop, and Tiziana Margaria. The fixpoint-analysis machine. In *CONCUR '95: Concurrency Theory, 6th International Conference, Philadelphia, PA, USA, August 21-24, 1995, Proceedings*, pages 72–87, 1995.
71. Bernhard Steffen, Falk Howar, and Malte Isberner. Active automata learning: From DFAs to interface programs and beyond. In Jeffrey Heinz, Colin de la Higuera, and Tim Oates, editors, *Proceedings of the Eleventh International Conference on Grammatical Inference, ICGI 2012, University of Maryland, College Park, USA, September 5-8, 2012*, volume 21 of *JMLR Proceedings*, pages 195–209. JMLR.org, 2012.
72. Bernhard Steffen, Falk Howar, and Maik Merten. Introduction to active automata learning from a practical perspective. In *Formal Methods for Eternal Networked Software Systems*, pages 256–296. Springer, 2011.
73. Bernhard Steffen and Stefan Naujokat. Archimedean points: the essence for mastering change. In *Transactions on Foundations for Mastering Change I*, pages 22–46. Springer, 2016.
74. Joseph E. Stoy. *Denotational Semantics: The Scott-Strachey Approach to Programming Language Theory*. MIT Press, Cambridge, MA, USA, 1977.
75. Daniel Suarez. *Daemon*. Dutton Adult, 2006.
76. Daniel Suarez. *Freedom*. Dutton Adult, 2010.
77. S. Tadelis. *Game Theory: An Introduction*. Princeton University Press, 2013.
78. Gerald Teschl and Susanne Teschl. *Mathematik für Informatiker. Diskrete Mathematik und Lineare Algebra*. Springer, 2010.
79. Ronald J. Tocci. *Digital Systems: Principles and Applications (5th Ed.)*. Prentice-Hall, Upper Saddle River, NJ, USA, 1991.
80. W. F. Truszkowski, M. G. Hinchey, J. L. Rash, and C. A. Rouff. Autonomous and Autonomic Systems: A Paradigm for Future Space Exploration Missions. *IEEE Transactions on Systems, Man, and Cybernetics, Part C: Applications and Reviews*, 36:279–291, 2006.
81. Matthew Turk and Alex Pentland. Eigenfaces for recognition. *J. Cognitive Neuroscience*, 3(1):71–86, January 1991.
82. Arie van Deursen, Paul Klint, and Joost Visser. Domain-specific languages: An annotated bibliography. *SIGPLAN Not.*, 35(6):26–36, June 2000.

83. Moshe Y. Vardi. Automata-theoretic model checking revisited. In Byron Cook and Andreas Podelski, editors, *Verification, Model Checking, and Abstract Interpretation, 8th International Conference, VMCAI 2007, Nice, France, January 14-16, 2007, Proceedings*, volume 4349 of *Lecture Notes in Computer Science*, pages 137–150. Springer, 2007.
84. Moshe Y. Vardi. Model checking as a reachability problem. In Olivier Bournez and Igor Potapov, editors, *Reachability Problems, 3rd International Workshop, RP 2009, Palaiseau, France, September 23-25, 2009. Proceedings*, volume 5797 of *Lecture Notes in Computer Science*, page 35. Springer, 2009.
85. John Vince. *Foundation Mathematics for Computer Science: A Visual Approach*. Springer, 2015.

Index

© Springer International Publishing AG, part of Springer Nature 2018
B. Steffen et al., *Mathematical Foundations of Advanced Informatics*,
https://doi.org/10.1007/978-3-319-68397-3

Printed by Printforce, the Netherlands